Palgrave Studies in Prisons and Penology

This is a unique and innovative series, the first of its kind dedicated entirely to prison scholarship. At a historical point in which the prison population has reached an all-time high, the series seeks to analyse the form, nature and consequences of incarceration and related forms of punishment. Palgrave Studies in Prisons and Penology provides an important forum for burgeoning prison research across the world.



More information about this series at
http://www.palgrave.com/gp/series/14596

Diete Humblet

# The Older Prisoner

palgrave
macmillan

Diete Humblet
Department of Criminology
Vrije Universiteit Brussel
Brussels, Belgium

Research Centre on Care for Older Adults
Odisee University College of Applied Sciences
Brussels, Belgium

Palgrave Studies in Prisons and Penology
ISBN 978-3-030-60119-5 ISBN 978-3-030-60120-1 (eBook)
https://doi.org/10.1007/978-3-030-60120-1



Cover illustration: Ilona Kryzhanivska / Alamy Stock Photo

This Palgrave Macmillan imprint is published by the registered company Springer Nature Switzerland AG.
The registered company address is: Gewerbestrasse 11, 6330 Cham, Switzerland

*For Hilde and Romain†*
*Who gave me life and moral compass*

# Foreword

Diete Humblet's impressive book is the first result of an interdisciplinary research programme aiming at analysing the dynamics of relations of (extreme) dependency in different institutions and to understand the experiences of loss and recovery of dignity, autonomy, and agency. And older prisoners are indeed an important illustration of such situations of *extreme institutional dependency*.

Representing still a small group within most European prison populations, their numbers are growing—but research on their situation is rare. In 2005, Elaine Crawley and Richard Sparks coined the term 'institutional thoughtlessness' to describe the lack of attention to their specific prison pains and needs in the English prison system. International prison research has since slowly started describing their double plight resulting from the combination of old age and deprivation of liberty. But Diete Humblet's book adds a new and innovative approach to the topic. Starting from the research question *what is distinctive about (the experiences of) older prisoners?*, she quickly came to the conclusion that more gerontological knowledge was needed to fully understand the needs and experiences of older persons, whether in prison or not. For the research question has two sides to the same coin: to what extent are older prisoners' experiences similar or different from *other prisoners* but also from *other older adults* in outside society? So, on top of her double degree in law and criminology, she engaged in an interdisciplinary Master in Gerontology. She then spent many months doing ethnographic research in two Belgian prison settings, comparing experiences of older prisoners in a traditional 'integrated' prison regime and in a special 'segregated' unit, combining shadowing, observation, informal conversations, face-to-face, and postal interviews. Even moving to another part of the country in order to be able to immerse herself in daily prison life as it is experienced by those living and working there—from six o'clock in the morning to late at night.

The result is a fascinating integration of the penological focus on the double vulnerability of older persons in prison, the impact of the prison environment, their coping mechanisms, and the gerontological emphasis on the different dimensions of age and the potential remaining positive and enriching aspects of ageing such as meaning in life and agency. Rejecting the predominant *deficit model* of ageing, growing older is construed here as an existential rather than simply a pathological process. Her research illustrates the importance of acknowledging not only the variability in pathways of older prisoners to *prison* but also to *ageing*. She describes the many faces of 'institutional thoughtlessness' towards older prisoners, even in regimes meant to counter its detrimental consequences, resulting from the negation of the different dimensions of age of her participants. And how the prison system could easily account for their continued search for meaning in life and generativity needs—which are 'commonly unacknowledged, unrecognised, and certainly unaddressed'.

This research is not only the result of an academic interest but also of the passionate involvement of Diete with her topic. Her sensitiveness towards her research subjects is remarkable. She kindly but firmly corrected my all too often references to 'elderly' prisoners as a form of 'ageism'—the (unintentional) stereotyping and reduction of persons to only one characteristic—age. She has since used the results of her research to write a white paper aiming at impacting Belgian prison policy and practice and is currently developing a training programme for prison staff in order to raise their awareness of this all too forgotten population. But the relevance of her research extends far beyond Belgian borders.

It has been a privilege to be her supervisor.

Brussels, Belgium
August 2020

Sonja Snacken

# Acknowledgements

This book reports the results of a PhD research project that commenced in 2013 and became the point of departure to pursue my own line of research, which is situated at the crossroads of criminology, penology, and gerontology.

This fundamental scientific research project would not have been carried out without the financial support of the Research Foundation Flanders (FWO). I am therefore grateful to this external financial contributor and thankful for their support.

Furthermore, I felt very fortunate to work at the Vrije Universiteit Brussel in Belgium, a university that is renowned for its strong humanistic roots and DNA. The University and the research Group Crime & Society allowed and encouraged me to conduct my research in a critical, independent, and free-inquiring way yet with respect for humanistic values.

I would also like to express my gratitude to the Belgian Prison Service as well as the local prisons, for enabling me to conduct empirical research within the penal realm. It is commendable that this prison system has gone through great lengths to make its walls permeable to the research community. Against this background, I consider myself as a privileged witness.

The book gives voice to all those who shall remain anonymous for the reader. A special word of thanks goes out to all the prisoners, all the prison staff members, and all other actors who I have encountered in the two prison settings where I conducted my fieldwork. I feel both privileged and honoured that you have allowed me to share your space, thoughts, and behaviour, which lie at the heart of this book.

In view of the foregoing, I would also like to take this opportunity to thank the editors and editing assistant at Palgrave Macmillan for their support in publishing this book and for having provided me with the opportunity to share this piece of work.

I am also grateful for the help of Fraser King for his diligent proofreading of this manuscript. A heartfelt thanks goes out to Dirk van Zyl Smit, Azrini Wahidin, Liesbeth De Donder, and Johan Bilsen for sharing their expertise and their constructive criticism of the research.

Lastly, I am also particularly indebted to my supervisor, Prof. Dr. Sonja Snacken, for guiding me throughout my research journey and leaving her mark on my life as well as my scientific work. I celebrate you for having kept me under your wings until I was able to take flight. I will yet forever remain one of your little birds.

Brussels Diete Humblet
July 2020

# Contents

# 1

# Introduction

## Introduction

Why should anyone be interested in the experiences of those who are spending the latter, or last stages of their lives in prison? As a prison scholar with a long-standing scientific interest in the older prisoner, I have found myself pondering that same question. Even after many years studying imprisonment in later life, I still find myself accounting for the study of the processes and problems of ageing in conjunction with the study of punishment of crime and prison management. This, in itself, says a great deal about our ideas and understanding of imprisonment as well as later life. The first section of this introduction attempts to formulate an answer to this question through analyses of the factors that may have affected our thinking in terms of older adults' involvement in criminal behaviour and the way this is managed. The second section of this introduction is concerned with the scientific research project that forms the basis of this book and concludes with an overview of the research questions that are answered in the following chapters.

In the words of Otto Pollak, even "old criminals offer an ugly picture and it seems as if even scientists do not like to look at it for any



D. Humblet, *The Older Prisoner*, Palgrave Studies in Prisons and Penology,
https://doi.org/10.1007/978-3-030-60120-1_1

considerable amount of time" (Pollak 1941: 213, own emphasis). In this area of interest, where the disciplines criminology and gerontology meet, this is to be considered as one of the most cited phrases and most poignant observations. Pollak's early striking observation, that (the control of) deviant behaviour amongst older adults is neglected by the scientific community, has been supported by more recent scholarship. Powell and Wahidin (2006), for example, have already suggested that developing the study of ageing constitutes one of the significant challenges for contemporary criminology. The prolonged failure of mainstream criminology to give adequate attention to the phenomenon of ageing is at best questionable and at worst alarming. Not least because ageing populations *do* in fact increasingly permeate the field, yet more importantly, it is because the pre-eminent domain of criminologists is to inquire about that which is considered as 'deviant'.

Amongst other things, it is often assumed that this negligence has been dictated by the well-established relationship between old age and crime (see, e.g. Sapp 1989), the foundations of which were laid out by Adolphe Quételet, who belonged to the first generation of scientists to suggest a strong relationship between age and criminal behaviour. Quételet's (1984) age-related curve of crime (also known as the 'age-crime curve') showed that criminal behaviour strongly decreased after the age of 50. This was one of the first etiological studies into the explanations of crime at a later age that made use of a target group that was not only 'known' but also relatively easy to research: a group of (relatively older) perpetrators who had already been sent to prison. The study of criminal offending had given a slow but steady impetus to inquire about an older population in the prison setting. However, at the same time, these findings provided some evidence to strengthen the efforts of criminologists to further develop scholarship in view of young(er) adults.

Although 'elderly'[1] *offenders* had become the subjects of more extensive research in the 1970s, studies examining imprisonment and adjustment to imprisonment in later life remained scarce. Most researchers have continued to focus primarily on the effects of incarceration on young male prisoners, implying that older inmates, including older female prisoners, are of little concern. The lack of interest in prisoners of advanced age became noticed by a minority of researchers, who described older prisoners in very apt terms: Ham (1976) dubbed them 'the forgotten minority', whereas Krajick (1979) made reference to 'forgotten people'. Goetting (1984) pointed out their low-priority status in society as a whole, as well as within the prison structure. She attributes this to the fact that in combination with their double minority status as being 'older' and 'criminal', older adults constitute a relatively small proportion of the prisoner population.

Over time, greater attention has been paid to changes within the prison(er) population(s), to some extent prompted by prison overcrowding, and in particular to the evolution in the numbers of 'older prisoners' and their pressing health needs. This is paralleled by societal and scientific evolutions that have come to recognise prisoners and older adults as more vulnerable groups of persons within society (Cuéllar et al. 2015; Martin et al. 2015). More recently, 'older prisoners' have been denoted as a 'doubly disadvantaged' group (Tarbuck 2001: 369) in terms of their greater morbidity and more complex health needs in comparison to younger prisoners and the general population of a similar age (Fazel et al. 2001). Over the last decade, the impact of the prison environment on older prisoners has received closer scrutiny in penological research, with Crawley and Sparks (2005) coining the term 'institutional thoughtlessness' to describe how older prisoners are disadvantaged by prison infrastructure, prison regime, and the prison staff culture (Crawley 2005). Accordingly, as a result of poor treatment and conditions, 'older prisoners' have been recognised as a group for whom imprisonment amounts to a 'double punishment' (PRT 2008). This has prompted, inter alia, a dialogue about establishing a 'geronto-penitentiary policy', including a discussion on whether older prisoners should be integrated or segregated from the general prison population in specialised units. As we will see later in this book, this discussion is hardly resolved in the existing body of literature. Rather, there appears to be an absence of well-grounded empirical and theoretical research scrutinising and comparing existing initiatives, which is addressed by this book.

Despite a growing attention for an older segment of the prisoner population, a full-fledged joint scientific interest in later life and imprisonment still seems far from established. To date, existing penological studies focussing on older prisoners tend to view ageing mostly through a negative and fixed lens as a result of which these individuals remain largely disempowered. Less traditional gerontological scholarship, by contrast, has learned to focus on aspects of ageing, which included 'positive' and enriching aspects such as coping, agency, and meaning in life. This duality is the common thread tying together the different chapters of this book. In order to gain a deeper understanding of the meaning of age(ing) in the prison realm, the limited body of existing literature—that has yet to incorporate many insights of gerontology—will be further developed with a view to an older prisoner population (cf. for a brief overview: Wahidin and Aday 2005; as well as other instances: Crawley and Sparks 2005; Doron 2007). Conversely, gerontologists in turn have tended to fail to investigate the nature of punishment of older adults' deviation from norms. From this it follows that, within other residential settings such as nursing homes, the most extreme and violent cases remain to fall between the cracks of the system and of our scientific attention. Whilst this topic falls outside the scope of this book, it provides food for thought and gives impetus to further advance the study of punishment and old age. The added value of this book is that it advances the frontiers of different disciplines by integrating a penological gaze, which translates into the desire to study the specific pains and vulnerabilities of older adults in prisons, with a gerontological gaze that examines gains for older prisoners and deconstructs the complexity and heterogeneity of their experiences by looking at many dimensions of age.

## Studying the Older Prisoner: The Belgian Case

What prompted this study? As in other Western countries, the number of prisoners in later life has risen sharply in Belgium in recent years: according to the annual penitentiary statistics of the Council of Europe (SPACE I Statistics 2007–2013; Aebi and Delgrande 2009, 2010, 2011a, 2011b, 2012, 2014a, b), the 'stock' of 'older' inmates (i.e. 60+) has increased annually with an average of around 12%. The ageing Belgian prisoner population has increased greatly in number over the past years: the prison population aged 60 or more has virtually doubled between 2007 and 2013. In this period, the absolute number of prisoners aged 60+ rose from 248 to 486. Although older prisoners still remained a minority in Belgian prisons (ranging from slightly under 2% for 65+ to 4% for 60+ of the total Belgian prison population), the numbers suggest that is still an ongoing trend, and as such, prisons as societal institutions should address current and future challenges in terms of dealing with more ageing prisoners.

Although the overall number of prisoners ('TPP') has also increased, the graph below indicates that the number of prisoners aged 60 or more ('EP') has soared (cf. table below). In comparison, from 2007 to 2013, the general Belgian population minus the total prison population (GP minus TPP) has grown with an average of 0.79% per year (Statistics Belgium, Statbel). Similarly, the total number of prisoners minus the elderly prisoner population ('TPP minus EP') annually increased with 4.04% on average, whereas the elderly prisoner population ('EP') increased with an average of 11.90% within the same period. In sum, the increase in the relative number of elderly prisoners is slightly under 50%, whereas, on average, their absolute numbers have increased with about 11% per year. This information has been presented in graph form below (Fig. 1.1):

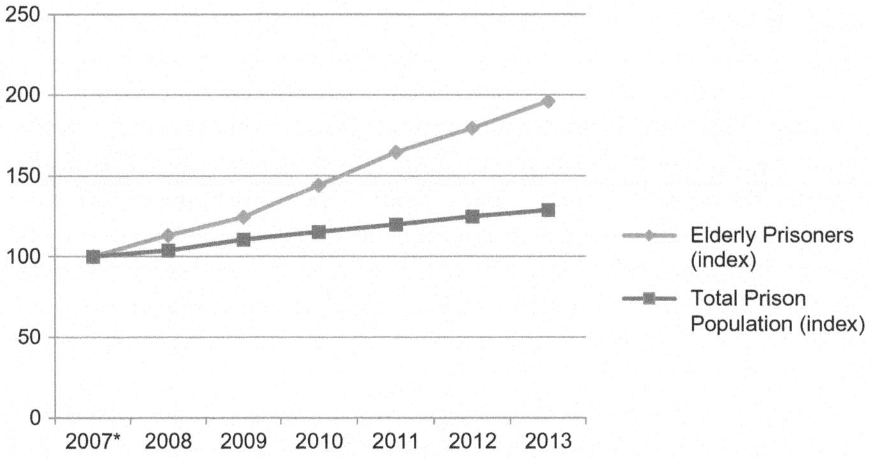


**Fig. 1.1** Evolution of the Belgian prison population (60+) between 2007 and 2013

Whilst it is likely that older inmates will remain a relative minority in most prisons, the underlying social and penal mechanisms behind this evolution suggest that their absolute numbers will continue to rise. Belgian prison policy has, to a certain extent, become more aware of the importance of this phenomenon and the specific challenges it can raise to existing prison regimes. This has resulted in the creation of two specialised units for 'elderly prisoners or disabled persons' (Mistiaen et al. 2017).[2] This presented a significant opportunity for comparative empirical research by conducting an ethnographic research into the meaning of imprisonment for older prisoners in two disparate Belgian prison settings: i.e. an integrated prison on the one hand and a more specialised unit on the other hand. Not only does this book reveal more about older prisoners in Belgium, it will also develop and illustrate a way of studying older prisoners that can be applied worldwide.

## The Fieldwork

The study that forms the basis for this book focussed on the meaning of imprisonment for the older prisoner.[3] The starting point of this book was the acceptance that there are multiple social realities; hence, multiple meanings are ascribed to the older prisoner's experience. Their search, which is underpinned by social constructivism, allows consideration of multiple realities. It thus attempts to document the meanings, experiences, and perceptions of the participants, considering that 'reality' cannot be merely constructed by the inquirer. Bearing this in mind, it seems more appropriate to consider this research as an exchange of dialogue between the researcher and the participants. The study of meaning for the older prisoner, considered through one's lived experiences, could only be captured by embedding oneself in the own environment of the older prisoner. This was accomplished by undertaking an *ethnographic study*, which has been adopted in the social sciences as:

> The study of groups of people in their natural setting, typically involving the researcher being present for extended periods of time in order to collect data systematically about their daily activities and the meanings they attach to them. (Noaks and Wincup 2004)

The study adopted a multi-method ethnographic approach, combining formal and informal conversations with multi-site participant observation and shadowing. The field was entered with "an open mind but not an empty head" (Lewis-Beck et al. 2003), and a reflexive stance was adopted throughout the entire study. The fieldwork was completed after spending a total of one year in two prison sites, both located in the Flemish Region of Belgium, where the highest concentrations of prisoners aged 65 years and older were housed. When studying such a 'hidden population' that is relatively small and is geographically dispersed throughout the prison system, we had to ensure that such an intensive long-term immersion in the field would yield a greater understanding of the topic. The prison settings were also selected for their exemplary value of the work presented, as enablers of building a knowledge base for integrating and segregating the older prisoner. I spent a total of eight months in the first prison setting (hereinafter denoted as 'P1'), which comprised one of the largest prisons in Belgium, where older prisoners were integrated into the general prison population and a medical centre yet no specialist accommodation was available. In addition to this, I spent four months at the 'gero-medical unit' of a second prison (hereinafter denoted as 'P2'), where frail older and ill or disabled prisoners were segregated from other prisoners. This very succinct description of research settings is complemented by a thick description of both prison contexts, as provided in Chaps. 3 and 4.

In both field sites, every prisoner having reached the chronological age of 65[4] at the time of the fieldwork was sought to participate in the study.[5] As a result of my long immersion in the day-to-day prison life, I had the opportunity to fully and intensively engage with a total of twenty older prisoners, mainly men.[6] Following the suggestion of O'Reilly (2009), I made efforts to talk with anyone and everyone, of all types, personalities, and roles in the field. As a result, I quickly came to learn that also younger prisoners, prison visitors, and prison staff members all provided a valuable source of information and provided meaningful insights during fieldwork conversations and observations. It seems more accurate, then, to describe this as an 'ongoing sampling' process (O'Reilly 2009). As a result, prison officers and nurses and some 'younger' prisoners became de facto included in the ethnography as well. Data gathered from observing and conversing with these actors were incorporated in the analysis to the extent that it contributed to a greater understanding of the meanings attached to imprisonment in later life, e.g. via interactions with the older prisoner, or lack thereof. It should be noted that prison staff placed greater emphasis on the 'deficit model' of ageing (Venegas 2019), by viewing old age mostly in terms of loss and decline. This offered little scope for examining the strengths and different dimensions of the ageing prisoner from their point of view. Meaningful informed consent was obtained relationally and was viewed as part of the ongoing close relationship that was developed between the research subjects and the researcher (Kendall and Halliday 2014). Against the backdrop of our research philosophy, there was no a priori exclusion of people with cognitive deficits from participating at the ethnographic interviews, although it was occasionally frowned upon in the field. The common assumption that people diagnosed with (symptoms of) dementia are 'empty shells' (Hellström et al. 2007: 608; Moore and Hollett 2003) and therefore could not produce any useful nor meaningful answers was as such strongly reflected by the position of certain prison staff members as well as fellow prisoners vis-à-vis certain research subjects with frailty. I tried to establish a safe context by, inter alia, engaging in conversations in a surrounding that was familiar and comfortable to the participant (Beuscher and Grando 2009: 3). Following on the conversations, Clarke and Keady (2002: 38) suggest that a continuing interest in the person as an individual after (the more formal part of) the interview has finished is essential if a lasting positive impression is to be created, and an instrumental 'hit-and-run' approach should be avoided at any rate (Hellström et al. 2007: 612). My time in the field was used to establish good relationships, build rapport, and foster relationships based on trust, warmth, and empathy, as a prerequisite to reduce power inequalities (Hellström et al. 2007) and to improve accessibility to the researcher.

Depending on the cognitive, the social, and the physical abilities of the participants and the opportunities created by the particular environments under study, more focus was placed either on the observational (in P2) or on the verbal (P1) methods of collecting data. Triangulation in this ethnographic study mainly refers to the combined methods of gathering data, such as participant observation, shadowing, and ethnographic interviewing, between which I shifted across time, place, and situation. Miscellaneous in-depth interviews with the key or core participants (i.e. a total of 20 older prisoners for both P1 and P2) and a handful of gatekeepers (few younger prisoner-servants and prison staff members) were conducted and tape-recorded. The fieldwork entails a total of twelve months, counted at an average of three full days of attendance a week, of participant observation and shadowing and draws mostly from impromptu in situ communication with more than 20 prisoners, more than 30 prison officers, and around 10 prison nurses. An open and undisguised form of participant observation, one of the main methods employed in the study, was employed to partially immerse myself in the field sites. Observing, and taking part in a range of activities, from the mundane aspects of daily prison life to more exceptional episodes has enabled a deeper understanding of older prisoners within the context of daily prison life and prison culture. To a smaller extent, participants were also subjected to the shadowing technique, as illustrated by the following excerpt taken from the field notes:

> Gerald [one of the older prisoners] proudly tells everyone about me that I am 'his shadow'. (Field notes P1, April 2014)

Shadowing allowed me to also experience, see, and feel things as much as possible from the perspective of the participant. Being neither imprisoned nor professionally involved with working prison life, I have never been a full participant or fixed aspect of prison life. In hindsight, my presence in the field can be best described as 'the accepted stranger' (O'Reill 2009: 176).

Tolich (2004: 101) has used the metaphor of an iceberg to help us in our understanding of confidentiality. The tip above the surface, he argues, relates to what is covered by the traditional notion of confidentiality (which he terms external confidentiality), i.e. ensuring that the participant remains anonymous in the final report (Tolich 2004: 101). In our study, we have attended to external confidentiality by acknowledging that we had knowledge on what the informants (regardless of their capacity) had disclosed but promised not to identify them in the final report of the research findings (Tolich 2004: 101). Furthermore, confidentiality was safeguarded during the fieldwork by protecting and not disclosing what each insider said in a private interview vis-à-vis other insiders as well as outsiders (with the sole exception of the supervisor, cf. 'peer debriefing'). Below the surface then lies internal confidentiality, i.e. the ability for research subjects involved in the study to recognise each other in the final publication (Tolich 2004: 101). In contrast, internal confidentiality, which is also generally known as 'deductive disclosure' in specialist literature within the domain of qualitative research (Kaiser 2009: 1632; Tolich 2004: 101–106), covers the identification of an individual's identity or group in a research report by using known characteristics, traits, or specifics of that individual or group. It was not until after the research process, notably during the stage of the report writing, that we were faced with the issue of internal confidentiality. I decided to use fictive names and identities in the book, which brings to life the informants without reducing them to numbers or abstract terms.

Making sense of the data constituted an eclectic process using a combination of strategies of analyses, such as axial coding, descriptive analysis, pattern analysis, narrative analysis, value analysis, and the constant comparison method, a strategy that dovetails with what has been termed as 'thick analysis' (Van Staa and Evers 2016).

## Structure Outline

The book is made up of seven chapters (including the introduction and the conclusion) wherein penological insights are integrated with gerontological insights, as the 'older prisoner' is subjected to both the experience of imprisonment and of growing older. The key research question that has driven the research project 'What is distinctive about (the experiences of) the older prisoner?' has been subdivided into three topical questions that run central to the book. This results in the following structure:

Chapter 2 investigates our understanding of 'the older prisoner' and lays a foundation for understanding the remainder of the book. It examines 'the older prisoner' as a constructed category that is accompanied with distinct characteristics and specific knowledge determining the way we think about, scientifically approach, and respond to this category. In this second chapter, the reader becomes familiar with an interdisciplinary framework, underpinned by some empirical data on the older prisoner, which is further developed in the following chapters.

Chapters 3 and 4 are centralised around older prisoners' experiences of daily life in prison. Both chapters investigate in what way the prison environment is shaped for the older prisoner and how this is experienced. The underlying notion of 'experience' refers here to older prisoners' lived experiences in prison and recognises the state of being confined in, and to, a prison environment as a particular type of experience ('Erlebnis') (van Manen 2016). These chapters pay attention to place, context, and actors. By describing the research settings in greater detail, this also captures part of the methodology. These chapters provide an analysis of the material prison world (Chap. 3) and the social prison world (Chap. 4) and its impact on the older prisoner, which is examined empirically for both an integrated and a segregated prison setting in Belgium.

Chapters 5 and 6 investigate how older prisoners integrate these experiences into their lives and give meaning to their lives. Both chapters deal with empirical data on how older prisoners face the combined experience of imprisonment and that of growing older. This is further broken down into their experiences of coping (Chap. 5) and their patterns of generativity (Chap. 6). These chapters draw on older prisoners' life experiences ('Lebenserfahrungen'), which are more inclusive than lived experiences ('Erlebnisse'), as life experience ('Erfahrung') is the accumulation of lived experiences and the understandings and sense that can be made of these experiences (van Manen 2016).

Chapter 7, the conclusion, summarises the core message of the book and discusses the broader implications of the findings. It highlights aspects that are useful for policymakers, scholars, and all those interested in learning about imprisonment in later life.

# Notes

1. Terms such as 'elders' or 'the elderly' to denote adults with an advanced chronological age are considered as pejorative and reductionist (Falconer and O'Neill 2007; Wahidin and Aday 2005). Where these terms are used, they have been taken directly from the relevant, mostly antiquated, literature. Whilst some recent studies on prisoners in later life still employ this term, I have sought to avoid these terms, where possible. In this work, the following terms or descriptors have been used interchangeably: 'old(er)', 'age(ing)', 'aged', 'of advanced age', and 'in late(r) life'.
2. See also the parliamentary question of Mrs. De wit (2010) addressing "the ageing population in Belgian prisons" (No. 1336) to the Minister of Justice. In the respective answer to this question, given by the State Secretary, the establishment of adapted celling for wheelchair-bound prisoners, the medical departments of three prisons, and the two specialised prison units 'for older prisoners' have been put forward as part of an adequate approach to an ageing prisoner population in Belgium.
3. The research project was funded by the Research Foundation Flanders (FWO). A very detailed account of the methodology and philosophy underpinning the research project is recorded in the unpublished doctoral dissertation of the author.
4. Despite 'older' prisoners being a priori delineated as 'having achieved the chronological age of 65 years', the book considers the different dimensions of age.
5. In all but two cases, the sample size of prisoner-participants tallies with the total number of research population within the sampled settings. Two older prisoners (respectively, housed in the first and second prison setting) refused to give their consent to be interviewed and were therefore not included in the study.
6. Of the 20 older prisoners involved in the ethnographic study, 15 prisoners were housed in the first studied prison setting (P1), whilst 5 resided at the specialised unit (P2). Sex offenders were overrepresented amongst older prisoners in the second setting (P2). The subset of female participants, with a total of three older female prisoners, was housed in a separate wing in the first prison setting (P1) and was characterised by the same strong heterogeneity as the subset of older male prisoners.

## References

Aebi, M. F., & Delgrande, N. (2009). *Council of Europe Annual Penal Statistics SPACE I—Prison Populations Survey 2007*. University of Lausanne and Council of Europe. Retrieved from http://wp.unil.ch/space/files/2011/02/SPACE-1_2007_English_rev.pdf.

Aebi, M. F., & Delgrande, N. (2010). *Council of Europe Annual Penal Statistics SPACE I—Prison Populations Survey 2008*. University of Lausanne and Council of Europe. Retrieved from http://wp.unil.ch/space/files/2011/02/SPACE-1_2008_English.pdf.

Aebi, M. F., & Delgrande, N. (2011a). *Council of Europe Annual Penal Statistics SPACE I—Prison Populations Survey 2009*. University of Lausanne and Council of Europe. Retrieved from http://wp.unil.ch/space/files/2011/02/SPACE-1_2009_English2.pdf.

Aebi, M. F., & Delgrande, N. (2011b). *Council of Europe Annual Penal Statistics SPACE I: Survey 2009* (p. 111). Council of Europe.

Aebi, M. F., & Delgrande, N. (2012). *Council of Europe Annual Penal Statistics SPACE I—Prison Populations Survey 2010*. University of Lausanne and Council of Europe. Retrieved from http://wp.unil.ch/space/files/2014/04/SPACE1_2010_English.pdf.

Aebi, M. F., & Delgrande, N. (2014a). *Council of Europe Annual Penal Statistics SPACE I—Prison Populations Survey 2012*. University of Lausanne and Council of Europe. Retrieved from http://wp.unil.ch/space/files/2014/05/Council-of-Europe_SPACE-I-2012-E_Final_140507.pdf.

Aebi, M. F., & Delgrande, N. (2014b). *Council of Europe Annual Penal Statistics SPACE I—Prison Populations Survey 2013*. University of Lausanne and Council of Europe. Retrieved from http://wp.unil.ch/space/files/2015/02/SPACE-I-2013-English.pdf.

Beuscher, L., & Grando, V. T. (2009). Challenges in Conducting Qualitative Research with Persons with Dementia. *Research in Gerontological Nursing, 2*(1), 6–11. https://doi.org/10.3928/19404921-20090101-04.

Clarke, C. L., & Keady, J. (2002). Getting Down to Brass Tacks: A Discussion of Data Collection with People with Dementia. In H. Wilkinson (Ed.), *The Perspectives of People with Dementia: Research Methods and Motivations* (pp. 25–46). Jessica Kingsley Publishers.

Crawley, E. (2005). Institutional Thoughtlessness in Prisons and its Impacts on the Day-to-Day Lives of Elderly Men. *Journal of Contemporary Justice, 21*(4), 350–363.

Crawley, E., & Sparks, R. (2005). Hidden Injuries? Researching the Experiences of Older Men in English Prisons. *The Howard Journal, 44*(4), 345–356.

Cuéllar, A. F., Tortosa, M. C., Dreckmann, K., Markov, D., & Doichinova, M. (2015). *Vulnerable Groups of Prisoners: A Handbook*. Center for the Study of Democracy. Retrieved fromhttp://www.ub.edu/ospdh/sites/default/files/documents/Handbook.%20VGP_EN__FINAL_1.pdf.

Doron, I. (2007). Heaven or Hell? Aging Behind Bars in Israel. *Hallym International Journal of Aging, 9*(2), 145–159.

Falconer, M., & O'Neill, D. (2007). Out with "the Old," Elderly, and Aged. *BMJ: British Medical Journal, 334*(7588), 316. https://doi.org/10.1136/bmj.39111.694884.94.

Fazel, S., Hope, T., O'Donnell, I., Piper, M., & Jacoby, R. (2001). Health of Elderly Male Prisoners: Worse Than the General Population, Worse Than Younger Prisoners. *Age and Ageing, 30*(5), 403–407.

Goetting, A. (1984). The Elderly in Prison: A Profile. *Criminal Justice Review, 9*(2), 14–24. https://doi.org/10.1177/073401688400900203.

Ham, J. N. (1976). *The Forgotten Minority: An Exploration of Long-term Institutionalized Aged and Aging Male Prison Inmates*. University of Michigan.

Hellström, I., Nolan, M., Nordenfelt, L., & Lundh, U. (2007). Ethical and Methodological Issues in Interviewing Persons With Dementia. *Nursing Ethics, 14*(5), 608–619. https://doi.org/10.1177/0969733007080206.

Kaiser, K. (2009). Protecting Respondent Confidentiality in Qualitative Research. *Qualitative health research, 19*(11), 1632–1641.

Kendall, S., & Halliday, L. E. (2014). Undertaking ethical qualitative research in public health: Are current ethical processes sufficient? *Australian and New Zealand Journal of Public Health, 38*(4), 306–310.

Krajick, K. (1979). Growing Old in Prison. *Corrections Magazine, 5*(1), 33–46.

Lewis-Beck, M., Bryman, A. E., & Liao, T. F. (2003). *The SAGE Encyclopedia of Social Science Research Methods*. SAGE Publications.

Martin, C., Rodríguez-Pinzón, D., & Brown, B. (2015). *Human Rights of Older People: Universal and Regional Legal Perspectives*. Springer.

Mistiaen, P., Dauvrin, M., Eyssen, M., Roberfroid, D., San Miguel, L., & Vinck, I. (2017). *Health Care in Belgian Prisons. Current situation and Scenarios for the Future. Health Services Research (HSR).* KCE Reports 293Cs. D/2017/10.273/65. Belgian Health Care Knowledge Centre (KCE). Retrieved from https://kce.fgov.be/sites/default/files/atoms/files/KCE_293Cs_Prisons_health_care_Synthese.pdf

Moore, T. F., & Hollett, J. (2003). Giving Voice to Persons Living with Dementia: The Researcher's Opportunities and Challenges. *Nursing Science Quarterly, 16*(2), 163–167. https://doi.org/10.1177/0894318403251793251793.

Noaks, L., & Wincup, E. (2004). *Criminological Research: Understanding Qualitative Methods*. SAGE.

O'Reilly, K. (2009). *Key Concepts in Ethnography*. London: SAGE Publications.

Pollak, O. (1941). The Criminality of Old Age. *Journal of Criminal Psychopathology, 3*, 213–235.

Powell, J., & Wahidin, A. (2006). Rethinking Criminology: The Case of 'Ageing' Studies. In A. Wahidin & M. Cain (Eds.), *Ageing, Crime and Society* (pp. 17–34). Willan Publishing.

PRT. (2008). *Doing Time: The Experiences and Needs of Older People in Prison*. Prison Reform Trust.

Question asked by Mrs. Sophie De Wit to the Minister of Justice about 'the Greying in Belgian prisons', V5301336, Kamer Commissie voor de Justitie, 2E ZITTING VAN DE 53E ZITTINGSPERIODE, CRABV 53 COM 056 1336. (2010). Retrieved from http://www.dekamer.be/doc/CCRI/pdf/53/ic056.pdf

Quételet, A. (1984). *Adolphe Quetelet's Research on the Propensity for Crime at Different Ages* (S. F. Sylvester, Trans.). Anderson Publishing Co.

Sapp, A. D. (1989). Arrests for Major Crimes: Trends and Patterns for Elderly Offenders. *Journal of Offender Counseling, Services & Rehabilitation, 13*(2), 19–44.

Tarbuck, A. (2001). Health of Elderly Prisoners. *Age and Ageing, 30*(5), 369–370.

Tolich, M. (2004). Internal Confidentiality: When Confidentiality Assurances Fail Relational Informants. *Qualitative Sociology, 27*(1), 101–106.

van Manen, M. (2016). *Researching Lived Experience* (2nd Edition). Routledge.

Van Staa, A., & Evers, J. (2016). 'Thick Analysis': Strategie om de kwaliteit van kwalitatieve data-analyse te verhogen. *Kwalon, 15*(1), 5–12.

Venegas, B. M. C. (2019). *Changing Age and Career Concepts in the Austrian Banking Industry: A Case Study of Middle-Aged Non-managerial Employees and Managers*. Springer Nature.

Wahidin, A., & Aday, R. H. (2005). The Needs of Older Men and Women in the Criminal Justice System: An International Perspective. *Prison Service Journal, 160*, 13–22.

# 2

# The Construction of the 'Older Prisoner'

## Deconstructing the 'Older Prisoner'

Much ink has flown on establishing a clear definition of what constitutes an older prisoner, yet much less attention has been paid to the meaning that is attached to the 'older prisoner' category and how it is socially constructed. How we have begun to conceive and how we talk about some people in terms of 'older prisoners' today reveals that this group is considered deviant in two ways. On the one hand, these are individuals who are believed to be involved in committing crime, which applies by definition to all those who have been imprisoned. On the other hand, the older prisoner also deviates from another norm, one that is determined by reference to 'the average prisoner'. From this viewpoint, the older prisoner is conceived as a *social problem* that must be *managed*. The idea of managing social problems through segmentation is not a novel one yet is here—as we will discuss later in this chapter—once more applied to the older prisoner who is placed in separate locations.



D. Humblet, *The Older Prisoner*, Palgrave Studies in Prisons and Penology,
https://doi.org/10.1007/978-3-030-60120-1_2

# Meanings of Old Age

The concept of *age* is paramount in understanding why someone might be considered as an older adult. This poses questions as to what constitutes 'older' and on which basis one is labelled as such. Age, and by extension old age, is attributed such a wide range of meanings that it must be approached from a variety of perspectives. This chapter deals with the most common meanings of age whilst examining how this manifests itself in prison. In doing so, it lays the groundwork for a more detailed examination of the experiences of the older prisoner that will follow.

## Chronological Age

It should be noted that the concept of age is not entirely free of social structures, which is reflected in the way it is interpreted by various societies, systems, and cultures. Age(ing) has been predominantly tackled by gerontologists and has received only scant attention in criminological and prison scholarship, thus requiring closer scrutiny of its interpretations within this system of knowledge. This book draws on an essential distinction between chronological and other meanings of age.

Chronological age (also known as the 'calendar age') generally refers to the amount of *chronological time* that has passed since birth, which is often expressed in years. Chronological time *can* be measured, inter alia, by how many times the earth revolves around the sun.[1] It should be noted in this regard that the chronological dimension is closely related to the *legal definition* of age, defined as 'the chronological age at which a person is entitled to access specific goods, services or benefits' (Hayslip and Panek 1993: 5) or, more generally, the cut-off point at which certain rights or obligations can be retained in life. Chronological time—and therefore also age—has thus been 'institutionalised' in societies (Dannefer and Phillipson 2010: 373; Baars 2007: 2, emphasis added). Although the chronological dimension of age has become one of the most widespread and well-established interpretations, it has also been subject to a fair amount of criticism.[2] One of the main issues is that chronological age cannot be seen as a very good predictor for other aspects in an individual, such as the psychological, the social, and the biological aspects of the state of a person (Hayslip and Panek 1993: 5f.; Neugarten and Hagestad 1976: 36). Similarly, Dannefer and Phillipson (2010) also draw our attention to the fact that calendar age cannot account in itself for ageing nor developmental processes. Although the use of a chronological approach for including populations in research about ageing or 'the aged' is questioned (Dannefer and Phillipson 2010: 368), it can seemingly serve as a valid indicator when used in combination with other variables at a group level (cf. Baars 2009: 88). Bearing these concerns in mind, this still raises the question which timespan is targeted when constructing old age.

When looking at the discussion of the World Health Organization (hereafter 'WHO') on the plight of the lack of general agreement as to what constitutes an older person, it can be noted that most developed countries accept the chronological age of 65 years to define an old adult (Kowal and Dowd 2001). The United Nations generally refers '60+ years' to the older population (Kowal and Dowd 2001). The WHO discussion reflects the difficulties of defining the beginning of old age, whilst at the same time not only indicating the role of chronological time in most developed countries but stressing the increased importance of other *socially constructed meanings* (e.g. loss in roles). Although there is a lack of scientific consensus on the definition of old age, the WHO generally refers to the United Nations agreed cut-off age of 60+ to denote the older population. For developing countries, however, it proposes a working definition of 50 years and older to define the beginning of old age. Even the meaning of chronological age is thus subject to considerable change, dependent upon increased longevity as well as the social and the specific context.

A similar line of reasoning applies to the prison context, where there is a lack of consensus on how to define an 'older' prisoner based on chronological age. As Golden (1984: 3) witnessed, "there is no uniform, fully agreed-upon, chronological age at which one becomes 'elderly'." Many authors have attempted to provide cut-off ages to provide an answer to what constitutes an 'older' prisoner. At the beginning of the 1980s, Rubenstein (1982: 288) identified that the majority of studies use an age of 50 or 55 to denote an older inmate. The same author pointed out that studies ranged from a minimum cut-off age of 25 years up to a maximum of 82. Uzoaba (1998: 2) reported that the criminal justice system itself often denotes offenders in their mid-50s or over as 'older', whilst studies tend to employ higher thresholds, provided that enough data or cases were available (Newman et al. 1984). This has been confirmed by more recent studies (see Lemieux et al. 2002: 44). A review of recent literature (see, e.g. Fiselier 2007), showed that the age limits used in studies vary from a minimum chronological age of 50 up to 65 and more whilst sometimes employing a combination of several age cut-offs. A significant number of authors use the term 'old(er)' or any other similar term to denote prisoners aged at least 65 years (Humblet and Decorte 2013; Crawley and Sparks 2005; Flynn 1992; for 'older offenders' in the criminal justice system, see van Turenhout and van der Heijden 2009), whilst others find themselves at the other end of the scale and employ a threshold of 50 years, which is one of the lowest (Chu 2016: 6f.; Wahidin 2004: 26; Aday 1994: 48; Grant 1999: 1; Uzoaba 1998: 2f.). In addition, there exists a third group who occupy an intermediate position and introduce cut-off ages of 60 years (Fazel et al. 2004: 396; Howse 2003: 2; Allersma 1970: 69) or 55 years (Mann 2008: 79; Aday 2003; Sapp 1989: 20) to denote older prisoners. Maschi et al. (2012: 1) point to the cultural differences that exist between different countries as to the definition of an older adult prisoner. Research that stems from Australia and the United States generally considers prisoners aged 50 or over as part of this old-age category (Howse 2003), whereas the United Kingdom research on older prisoners is more inclined to accept higher ages, at 60 or 65 (Maschi et al. 2012: 1; United Nations Office on Drugs and Crime 2009: 123).

Howse (2003: 2) has suggested that the prison environment imposes specific age criteria that are unique to prison life due to the unrepresentativeness of the prison environment. Given that the population in prison is predominantly made up of young(er) adults, the prison population reflects a somewhat different age structure, as opposed to the general population. Howse (2003: 2) argues that prisoners start to think of themselves as 'old' at the age of 50 and, as a result, might belong to a class apart because of their age. Similarly, he suggests that this anomaly could also affect how prisoners are regarded by their younger counterparts (cf. others-perceived age) (Howse 2003: 2). We will return to these subjective dimensions of age later in this chapter. According to Wahidin and Aday

(2005), prison officers tend to use 50 or 55 to define an older offender, and this comprised the preferable entrance age for the older offender units in the United Kingdom. Belgian research on older prisoners, which was conducted from the point of view of social care workers and service providers working with (older) prisoners, showed that most respondents linked the idea of 'old' to chronological age but without any consensus (Maelstaf and Claessens 2016: 67). Maelstaf and Claessens (2016: 68) have furthermore clustered the motivations of these professionals to justify their choice in favour of a certain age criterion. They hereby distinguished four categories, of which two have been subsumed under the notion of 'instrumental' motivations (referencing to similar issues extramuros, or to one's own task or practical experience); and the other two have been denoted as 'prison-specific' motivations that stress the comparison with the outside (i.e. either by drawing on the average age of prisoners or on the idea of accelerated ageing amongst prisoners) (cf. infra) (Maelstaf and Claessens 2016: 68). These findings have borne witness to the myriad of visions surrounding the definition of an older prisoner. As already pointed out by Newman et al. (1984), the definition of who is an older adult is highly dependent on *the data* that are used and the *purpose* of classification.

In view of the foregoing, other types, indices, meanings, or dimensions have emerged as an alternative to, or extension of, the chronological age concept. The most common of these will be discussed in more detail in the following sections. As the measured time is purely instrumental and empty, according to Baars (2012: 11), it only gains meaning through the visions, perspectives, and stories that lie behind and make use of the chronological time perspective. Provided that old age is as loose, fluid, and subjective as is suggested in the literature, it is going to be more context-dependent, and our ethnography will define it.

## Biological Age

The biological age construct of an individual, sometimes referred to as physiological age, is comprised of two strands. First, it refers to the physical changes that a person undergoes over time. For example, the decline in physiological reserves, the condition of an individual's organ and body systems (Hayslip and Panek 1993: 6), or biological markers of ageing.[3] These include, amongst other things, the number of cells that can divide themselves and the state of the cardiovascular or nervous system (Hayslip and Panek 1993: 6). Although there is a debate about the use of such biomarkers of biological age, the emergence of grey hair, wrinkles, sensory deterioration, and so forth are often regarded as signs of ageing. A patient with Werner's syndrome, or a Progeria patient chronologically aged 6, but biologically aged 60, serves as a reminder of the discrepancy between one's chronological and biological ages (Hayslip and Panek 1993: 6). Conversely, a less extreme example is living either a very active or more sedentary lifestyle (Hayslip and Panek 1993: 7) that results in a chronological age higher or lower than the biological age suggests. Secondly, biological age can be expressed in terms of life expectancy, longevity, and mortality (Hayslip and Panek 1993: 7). Bearing this in mind is of great importance as links might be established to punishment—particularly with regard to custodial sentencing—which is related to and expressed in time (Foucault 1989: 318).

## Biological Condition

This section adopts a biological-medical lens to examine the old-age prisoner population that has been included in the ethnographic study. Our empirical data touching upon the biological condition or bodily processes of the older prisoners involved in the study primarily pertain to the presence of (self-)reported experienced biological processes, regardless of whether they could be regarded a result of diseases and damage ('secondary ageing'), or as part of the universal changes that occur in human biology ('primary ageing') (Pachana 2016: 23f.).

In the first prison setting, older prisoners have reported more acute diseases or ailments (e.g. bronchitis, gall crisis, cold, influenza, tinea pedis, i.e. a fungal infection at the foot which has been acquired by showering in common showers). It is striking that in both of the respective prison settings, participants have mentioned one or more of the following chronic diseases (more or less indicated in a descending order to the extent possible): chronic back problem, bone and joint problems, cardiovascular disorder (mostly chronic heart failure), diabetes, hypertension, dementia, paralysis, lung problems, cancer, rheumatism, allergies (e.g. gluten intolerance), kidney failure, and multiple sclerosis. Against this backdrop, a handful of participants even indicated co- or multimorbidity, which was found to be intertwined in praxis. Although problems with urinary and faecal incontinence are not mentioned by the participants themselves, the existence of such problems has been pointed out indirectly by other prison actors (staff and/or fellow prisoners) in both settings. Miscellaneous symptoms or complaints that are mentioned by the core participants across both penitentiary settings include falling on the ground, experiencing headache, pain, dizziness, fatigue, muscular weakness, drowsiness, and loss of consciousness. Moreover, ranging from mild to more severe, sensory impairments (e.g. hearing and/or visual impairment) are fairly prominently represented in the observations of, and conversations with, both prisoner populations. Although it may be inferred that there is a higher concentration of prisoners experiencing secondary ageing in the second prison setting, it should be noted that we have found similar biological profiles amongst older prisoners housed in the first prison setting. Our findings suggest that similar biological profiles could be identified. Older prisoners with disparate biological ages were found to be dispersed over both prison settings. Turning our attention to the other participants (i.e. chronological-wise 'younger' prisoners), their biological ages are at times equivalent (or even higher) to those of the core participants. It is important to note here that our data accord with the work of others that in the sense that the manner of progress of ageing processes varies greatly (Baars 2012: 10).

## Older Prisoners' Relative Position

With regard to the second meaning attached to the biological dimension of age, we draw upon the concept of the 'pain quotient', as introduced by O'Donnell (2014). Against this background, the different meaning that the prisoner's relative position carries in the temporal landscape of imprisonment is examined. Further building on this, the fifth chapter bears witness to how older prisoners attempt to manage this landscape. In this respect, some prisoners consider their remaining time as extended ('an extended time horizon') and/or undertake actions to increase their longevity.

An inherent feature of imprisonment is that it imposes certain time relationships on the prisoner (Farber 1944: 158). Whilst it has been Foucault's contention that "everyone pays the same price when he loses his freedom; even more than the fine, she is an egalitarian punishment" (Foucault 1975: 318), other scholars have advanced that the temporal orientation impacts the challenge in terms of making sense of imprisonment (O'Donnell 2014: 202). O'Donnell has introduced the 'pain quotient' (PQ), i.e. the hurdle that must be overcome to make life in prison bearable (p. 201). Its appraisal draws on the one hand on 'time to be served', which includes sentence length and point in sentence, and on the other hand on 'time to be lived', which takes account of chronological age and life expectancy of the prisoner (p. 201f.). The knowledge that 'time is running out' is what makes the experiences of older prisoners distinctive (Deaton et al. 2009; Crawley and Sparks 2005; Wahidin 2004: 109) whilst intensifying the pain quotient. It is moreover argued that the pain quotient is greatest for the prisoner "whose life expectancy is shorter than his prison time and for whom the absence of a non-prison future must be confronted" (O'Donnell 2014: 202). By virtue of advanced chronological age alone, for older prisoners, a large span of time has already passed. Most core participants in our study had already spent several years in prison and thus have had already overcome the first shock of imprisonment at the time of the fieldwork. Several prisoners still had to serve several years and/or had been in the process of gaining early release. One core participant had been given early release at the time of the research and had been relocated to a nursing home. Several core participants had been permanently cut off from a non-prison future: three of the core participants have passed away during, or in the aftermath of, the actual fieldwork. Whilst an objective appraisal of older prisoners' temporal orientation would suggest high(er) pain quotients, most core participants still viewed themselves in terms of a non-prison future, despite advanced age and prison time. As O'Donnell has suggested, "impending old age may bring the absence of a non-prison future stronger into focus, hence requiring significant coping strategies or adjustments whereby it is imperative to devise a strategy for negotiating the temporal landscape of imprisonment" (p. 202, own emphasis), which is evidenced later in this book.

## Accelerated Ageing

Some scholars have argued that prisoners are subjected to accelerated ageing and, accordingly, have biological ages equivalent to individuals in the community who are 10 to 15 years older (House of Commons Justice Committee 2013, HC 89-I: 13; Roberts 2015: 25; Loeb et al. 2008; Fazel et al. 2001). This age differential has been attributed to factors arising prior to, as well as during, incarceration (Maschi et al. 2012: 1). The notion of accelerated ageing amongst prisoners is however contested by some scholars. Spaulding et al. (2011: 485) argued that the common assertion that inmates are "10 or more years older" in health terms than similarly aged free persons is an oversimplification of the health of prisoners. According to their findings, the 'shift' in morbidity is dependent on the age of the prisoner and the amount of time considered for follow-up (Spaulding et al. 2011: 485).

As part of the written evidence accompanying the House of Commons Justice Committee Report on older prisoners, Crawley (HC 89-I: 9) also adopts a very different view and casts doubt on previous research. In the report, she argued that:

> In my own view, however, our retirement age for men (65 years) is the age at which most people, both within and outside the prison community are starting to 'feel old'. I do not accept the generally held view that prisoners are generally 10 years older, both physically and mentally, than their counterparts in wider society. This was not borne out during my research: rather people from all walks of life age differently. The 'older' prisoners I have interviewed in the period 2002 to the present have been aged between 65 years and 84 years. (House of Commons Justice Committee 2013, HC 89-I: 9; HC 89-II: ev w28, own emphasis)

The hypothesis that prisoners would experience 'accelerated ageing' was also touched upon by some of the prison staff members involved in my own study:

> A nurse says to me: 'That man you saw earlier, he's 69 years, but he actually looked good for his age, huh?' I reply: 'Yeah, he actually did'. The nurse says: 'Yes, well, they often look older. That's because they're in prison'. I ask: 'Or because of their life before prison?'. 'That as well', she responds. She reiterates that prisoners are ten years older as a result of imprisonment; that you immediately see it. I ask if she also observes it among prisoners serving brief sentences. She says: 'Oh, no. When they served long time'. [...] In answer to my question 'Why do you think this is? she replies: 'because of prison conditions'. She continues: 'Yes, because of the circumstances, huh. That's in fact detention harm'. 'First and foremost, very bad food, exercise, air, bad teeth'. I say that I sometimes notice that people are fairly 'older' but still look fairly 'young, whilst also observing the opposite. She says: 'Yeah, work and prison commissary, that goes a long way'. She also says: That depends how you cope with your punishment mentally'. 'If you can buy something from the canteen and you buy wisely and you take some care of yourself, unless you smoke cigarettes, then it's also bad for your health'. 'But if you can't buy anything...'. (Field notes P1, July, 2014)

Few participants felt that their health (biological age) had deteriorated as a result of the prison environment. In this respect, especially female prisoners have asserted that it felt like as if they had physically aged ten years in prison (which was disproportionate to the amount of years they had already spent in prison). On a more general note, prisoners have indicated that their eyesight had strongly declined, which they ascribed to the high levels of exposure to the artificial prison lighting (cf. Wahidin 2005: 9). Some have made no reference to the impact of imprisonment on their biological age. Others have touched implicitly on their biological age, by denouncing the lack of adequate health care provided in prison (cf. Humblet 2014). Consistent with the existing body of literature (Mann 2012: 47ff.), the accounts of the core participants regarding access to, and use of, prison health care have been frequently characterised by, inter alia, a wrong or late diagnosis, chaotic dispensation of medication, absence of sufficient resources and/or staff, lengthy waiting times, wrongfully prescribed medication or a lack of adequate medication, degeneration caused by infrequent check-ups, and even death of fellow prisoners attributed to a lack of concern by healthcare staff. However, experiences stating the contrary were also recorded. These were often formulated as 'they do what they can', or 'requested medical treatment is received on time', thus representing a more resigned attitude (Humblet 2014: 3). The treatment of older prisoners, in the prison system and by the prison actors (pertaining to both professionals of custody and care), will receive closer scrutiny in the following chapters.

As a result of our inquiry, we found that several of the older prisoners from our sample had fallen multiple times, which applied both to the first prison setting (see, e.g. Field notes P1, May, 2014) and the second setting (Interview Antonius; Miscellaneous Field notes, P2), which can also be associated with the functional performance in interaction with the environment. This brings us to the functional age.

## Functional Age

The plethora of definitions of functional age have in common that they are an index of one's level of capacities to function or perform well in a specific task or an index of tasks (Hayslip and Panek 1993: 10). Whilst functional age is often conceptualised as 'a fourth type of age in its own right' (Settersten and Mayer 1997: 240), it has also been conceived as a combination of the biological and psychological age (Sterns and McQuown 2018: 250), or as a sub-dimension of biological, psychological, as well as social age (Settersten and Mayer 1997: 240). Functional age can be contingent or relative. A prisoner's functional limitations and level of capacities interact with the opportunities to engage in prison activities (such as prison work and physical exercise) and to fit the requirements of the prison environment. To this end, we have employed a 'person-environment' point of view of the functional age concept. Others have also subscribed to this view by referring to *functional impairment* as "a complex interaction between the physical abilities of a person and his environment" (Williams et al. 2006: 5, own emphasis).

## Daily Life

In examining the older prisoner's day-to-day functioning, the specific nature of the prison environment has already been considered. Advancing traditional insights concerning 'activities of daily living' (ADLs), Williams et al. (2006) coined the term 'prison activities of daily living' (PADLs) in their study on the functional impairment in geriatric (55+) female prisoners in California state prisons. In addition to the traditional ADLs (e.g. bathing, eating, toileting, dressing, and transferring in/out of bed or chair), they have identified five physical activities that are unique to independent daily functioning in prison: i.e. dropping to the floor for alarms, standing for head count, getting to the dining hall, hearing orders from staff, and climbing on/off the top bunk bed (Williams et al. 2006: 2). These findings could only be extrapolated to a lesser extent to our research population and settings. It is unsurprising that the some of these PADLs (i.e. dropping to the floor and standing for head count) could not have been corroborated by our results, as these activities are foreign to the Belgian prison system and its regimes. The notion of PADLs is therefore revisited in the light of the study site and its current context. Moreover, Williams and her colleagues have found functional impairment to be much more common in PADLs' measures than in measures of ADLs and have suggested that people who are independent in the community might be regarded as functionally impaired in a prison environment (Williams et al. 2006: 5). In this view, it is liable that prison exacerbates functional impairment for its captives as a result of their participation in physically demanding activities and/or a lack of access to environmental modifications for, or a lack of formal assistance with, functional impairment (Williams et al. 2006: 4). Herein, the question rises as to the specificity of the prison environment in terms of prisoners' functionality. Conversely, it could also be argued that the prison environment might mask, or facilitate the hiddenness of, other functional impairments. On grounds of both strands, as outlined above, our results further support the idea of interdependence between the (prison) environment and the functional impairment of its inhabitants.

## Communication

Prison is characterised by its own bureaucratic modes of communication, which are to a great extent formalised and impersonal. For most prisons, the pre-printed paper forms or slips that prisoners have to fill ('rapportbriefjes' in Dutch) constitute one of the most important modes of communication whereby prisoners can request access to most of the internal and external services in prison.[4] These report slips have to be handed over to a prison officer who is, in turn, responsible for its further distribution. Several participants need help with filling in such notes, either due to illiteracy but, more frequently, due to a reduction in their physical and/or cognitive abilities. For these prisoners, communicating is a perilous undertaking in a setting where there is an intern communication system that presupposes and draws upon a certain standard as regards functionality of prisoners. Moreover, the inability to write, read, hear, understand, or speak entails that such prisoners are to a large extent cut off from this means of communication intra-muros and extra-muros and in relation to themselves (e.g. self-expression or self-reflection). Its value is notably illustrated in one of the following chapters, which is interspersed with instances whereby the functional age of prisoners—in terms of their performance on a particular task (i.e. communication)—is instructive.

It should be noted that a dichotomous interpretation of functional age would neglect the complex reality that is hidden behind it. For instance, we have observed that some of the prisoners who had been confronted with the inability to write still found themselves capable to perform certain minor tasks in this area (e.g. to fill in puzzles, and/or mark or cross numbers on a bingo card, and/or fill in the prison canteen list). Conversely, others would biologically still be able to perform certain tasks yet need support in doing so as a result of a disadvantage or decline in terms of their cognitive abilities. Kinaesthetic and visual impairments in relation to written communication were—to a certain extent—addressed by some material support, such as glasses, but more often imply intangible support.

R: I cannot write.
I: Would you allow another person to do it for you?

R: No … Ehm. Then I would have to mention things that I would rather not mention to these guys [prisoners]. […] I'm working on a course that allegedly would prepare myself to be transferred to a hospital [residential centre]. But that's a written course work and because I can't write, they have found someone from Judicial Social Welfare to whom I narrate, and that person writes it down for me. Cause that's a major handicap; not being able to write.

I: Has that been like that for a long time?

R: Ever since my 46th year. That was the first day, that I couldn't do that anymore. I couldn't walk anymore. That was also part of it. And then my face went a little… But that turned out fine.

I: And what do you do when you have to fill in those report slips [in Dutch 'rapportbriefje(s)']

R: [Nods]. You have to make do with what you've got.

I had observed some reticence of this prisoner in asking for assistance, hence the following question:

I: Do you sometimes ask them [prison officers] for help?

R: Yes. I certainly do. If something needs to be written. And I have never had a … refusal like 'well, can't you do that yourself'. I have never encountered that…

I: Do you find it difficult to ask them?

R: Ehm…

R: Difficult … not solely towards the chef [prison officer], but even outside as well. Because if you ask something, ehm, to write down or whatever, you immediately receive the predicate of 'someone who can't read or write; that you're illiterate'. Because that is usually associated with each other. So, I *do* tell them, simply to avoid that image. Because … that rankles me a little [laughs].

I: How exactly do you avoid that?

R: Then I'll say: 'Yeah, but I can read though', and then I'll show them that I can. But writing doesn't go anymore. That locomotion is…

(Interview Albert, February, 2016)

There is a large stigma attached to illiteracy as it is often equated with stupidity. It seems moreover that those who must make compromises in terms of physical conditions seem more inclined to attach importance to mental conditions (especially intelligence). Such participants demonstrated a great deal of effort to hide their inabilities and appear functionally able. Prisoners seek foremost the aid of the psychiatric nurse and/or (one of) the prisoner-servant(s) ('porter').[5] Somewhat as a last resort, these prisoners also turn to prison officers. Whilst the same principle applies to the prisoners residing at P2, communication is more direct, and person oriented as staff is more approachable.

Direct personal contact with prison actors, which refers here to oral communication, may also generate its own set of problems. Prisoners who are faced with failing auditory functioning encounter difficulties in hearing orders from prison officers (cf. Williams et al. (2012) who already warned for this), and/or what other actors are conveying (e.g. conversations with sentence implementation court or fellow prisoners) with poor communication and social exclusion from other prisoners as two important consequences. Additionally, overcompensating auditory impairments by turning up the volume or speaking loudly may cause problems in terms of noise, which might in turn constitute an extra potential source of conflict(s) in interactions with other inmates. A significant proportion of the population under study makes no reference to the experience, or shows no signs, of hearing loss. Some prisoners with hearing problems do not act on it, whilst a small group of those with hearing loss make use of instruments such as hearing aids, individual headphone.

> R: Now I have two sets of hearing aids and normally those batteries would last for over a week … fourteen days. But in my case, they only last for a week.
>
> I: Why is that?
>
> R: Well, because I wear them for such long periods of time. I wear them twice as long as someone else would normally wear them. […] She [the audiologist] said I wear them for at least seventeen, eighteen, nineteen hours in my ears. And then it's normal that they only last for about a week […].
>
> I: And can you do without them?
>
> R: I can… But then I have to ask every minute 'what?'. And especially when it regards the television… Ehm, then those guys [prisoners] say that I turn it on too loud. However, there are others who also hear and see poorly. But well … I myself still have very good eyesight. I have to put on

glasses to read though, but when it comes to seeing at a distance, I see much better than many people with glasses.

I: Are there things that they could adjust in that sense?

R: [Silence].

I: A separate screen or a larger font for instance?

R: Ehm, no. A larger font would be helpful, but well … that must be built into the television software. And that's not that obvious, huh?

(Interview Albert, February, 2016)

In most cases, old-age prisoners manage to make themselves intelligible—at *least at the most basic level*—and, mutatis mutandis, also manage to understand other prison actors. The extent to which sensitivity must be demonstrated must correspond to the vulnerability of persons (Baars 2012). By drawing on the work of Stans et al. (2013) we have identified 'communication vulnerable persons' amongst our sample, who are denoted as 'those who struggle to communicate in a particular environment' (Stans et al. 2013: 302, own emphasis). We may infer from the above that this certainly applies to persons confronted with inaudibility, illegibility, and to—even more dire and vulnerable—plights of unintelligibility. For a group of core participants, particularly those with less cognitive capabilities, it is exceptionally challenging to understand others and to express oneself clearly. Whilst prison staff has been found to occasionally hint at these challenges, e.g. 'He doesn't understand it all that well' (e.g. Field notes P1, November, 2014), or 'You won't be able to have a conversation with him' (e.g. Field notes P1, May, 2014), these issues have not seldom been left unaddressed. Moreover, suchlike prisoners might communicate rather through behaviour and body language, of which a large part is most likely lost. Such communication has been more discernable in the special unit: 'You know there's something wrong when he starts cleaning tables or when he starts walking very tilted' (60+ prisoner) (Field notes P2, February, 2016), or 'perhaps he shows an aggressive response out of fear' (i.e. a quote from a prisoner-servant at P2; Field notes P2, February, 2016). To acknowledge such modes of communication presupposes a certain sensitivity, that is, to a large extent, absent in the prison context. We will return to this issue in Chap. 4, which bears witness to distracting and more invisible modes of communication (cf. infra on 'medicalisation').

To alert prison officers, prisoners can press a button that is attached to the wall in the individual cell or in the communal area to notify prison officers: this communication system causes difficulties especially when prisoners are unable to reach or press the button (e.g. as a result of a fall on the ground or disorientation).

## Ambulation and Exercise

There were no bedridden patients amongst the older prisoner population under scrutiny, with a single exception of a female interned mentally ill prisoner (P1) who had become bedridden at one point during the research due to a cerebrovascular accident.

Although most participants housed in prison setting no. 1 were able to walk independently (i.e. without any assistance), there were few exceptions where a prisoner would need occasional support from a wheelchair and staff to bridge long walking distances and/or is assigned a walker. However, there are prisoners without assistance who still encounter difficulties in getting around the prison environment (e.g. walking up the stairs with books) and/or in the process of being transferred outside prison (e.g. transferring into and out of the prison transportation vehicle whilst handcuffed) (Mann 2008: 132; Wahidin 2004: 132).

Virtually all participants required some form of ambulation assistance in prison setting no. 2: one walking with assistance of a crutch and another with assistance of a walker, and two prisoners were wheelchair-bound. These wheelchair-bound individuals need help with transferring in and out of bed and remain the bulk of the daytime seated in their wheelchair. One prisoner who walked without assistance shuffles and regularly holds onto the wall needed support from a wheelchair and staff to push him outside the unit.

Analogous to the other non-chronological dimensions of age, it also remains important here to avoid dichotomous thinking. One should rather think of functionality in terms of a continuum, whereby different degrees of functioning exist that should consider the institutional context, circumstances, and the particular act that is concerned. For instance, some of the core participants have been found to be still able to ambulate independently (such as in the prison unit), though they became more functionally impaired under certain circumstances, such as in the case of long walking distances, stairs, or steps (Miscellaneous Field notes P1 and P2).

Most prisoners (P1) did not participate at the common prison walk during airing time, with little exceptions and/or some pointing to previous endeavours. There was one core participant who exercised in the gym amongst younger prisoners. One core participant obtained individual walks upon his request to avoid victimisation. Some participants pursued exercising in cell, with merely one putting it into practice. Virtually all sentenced female prisoners participate at the walk (except for the one occupied with two work shifts).

The older prisoners at prison setting no. 1 passed up the opportunity to walk with other prisoners during joint walks, despite the presence of a ramp for less mobile prisoners. Most (older) prisoners did take part in the sporadic grouped walks at the inner courtyard of the prison organised by the psychiatric nurse assigned to the gero-medical unit. Whilst this was found to be a popular activity, this did not meet the needs of two older prisoners. For these latter individuals, physical exercise was limited. The wheelchair-bound prisoners in prison setting no. 2 do visit a physiotherapist multiple times a week. One prisoner participated at gym organised for 50+ prisoners.

## Maintenance

Prisoners in prison setting no. 1 were expected to make their beds, maintain their prison cell, clean their cutlery, recycle their garbage, and provide their laundry. Whilst several prisoners found these tasks daunting (especially cell maintenance), a very small minority of the older prisoners was entirely unable to perform such housekeeping tasks. In this case, the prisoner-servant takes over these tasks (or a member of the healthcare team in case of interned prisoners). As for prison setting no. 2, the inhabitants of the special unit were not expected to perform suchlike tasks regardless of their instrumental functioning, as these were assigned to the prisoner-servant and structurally embedded into his remit.

The overwhelming majority of prisoners in prison setting no. 1 managed to bathe and toilet independently (with the application of handrails in the showers and toilet), with few exceptions whereby there had risen issues or suspicions of incontinence. In prison setting no. 1, no assistance was provided with dinner, with the sole exception of offering mixed food. Most participants in prison setting no. 2 required some assistance with bathing and dressing and need some help with eating (e.g. in the form of adapted cutlery, assistance with cutting food). Some of the core participants went to great lengths to increase their own independency; e.g. one core participant tried to put on his own socks by using tricks (Interview Albert, February, 2016).

## Prison Work

(Older) prisoners who were still involved in paid activities (de facto solely prisoners from prison setting no. 1) mostly performed manual labour in prison, which comprised the core of available prison work. Our findings into the functional age in terms of competence in performing prison work suggest that prison officers and older prisoners tend to associate 'old age' with a worse or slow(er) performance. This is evidenced by the following conversations with prison officers at the workplace:

> A female prison officers states: 'The elderly are a little bit slower', whereupon she laughs. She also says: 'It's because I myself am also a little bit older, that I say so, huh?' She laughs and looks at her chronologically younger colleague and me. (Field notes P1, May, 2014)

And:

> A female prison officer replies 'well, we treat everyone the same' to the question what it's like to work with older adults. She subsequently says about the only 65+ prisoner working at the respective workplace at the time of the data allocation: 'He's a little bit slower, though'. (Field notes P1, May, 2014)

Conversation between an older prisoner and prison officer:

> Russell makes a reference to himself whilst he notes in passing: 'You know that this old man is fast, huh, chef!'. (Field notes P1, April, 2014)
>
> Prison officer Bruce states that you actually wouldn't label Russell as 'one of the elderly'. This officer continues: 'At times, he's even fitter and faster than me'. 'He's very well-preserved'. (Field notes P1, August, 2014)

Conversation with older prisoner:

> Dorothy says that older adults are not slower. She says that's all nonsense that they would be slower and work less well. She says she still can bend forward and reach the ground with the palms of both her hands. She does however say that old age entails that you need more time to recuperate from an effort, but that you can still do the same. (Field notes P1, September, 2014)

During our ethnographic research, we established that several prisoners, older than 65 years of age, were physically and/or mentally unable to perform prison work. Unemployment problems and the lack of an adequate labour supply in prisons have already been raised (e.g. Goetting 1985; and see especially for specialised facilities and units: Aday 2003: 129ff.; 146). Moreover, the work that is available in prison is generally not adapted to old age or the associated possibilities and limitations of some prisoners (cf. the notion of 'institutional thoughtlessness'). For example, where prisoners are paid on a piece-by-piece basis (in the prison workshops), some respondents claim to be in a disadvantaged position with regard to their compensation for prison labour. For example, some of them would receive fewer pieces per hour. This underlines the importance of the work assigned to them ('young people get the better work') and having a young functional age. For example, a respondent describes the reaction of the prison staff as follows: 'If you can't do it, you can go back upstairs [to your cell]' (Interview Anita, September, 2014), with the understanding that they also no longer had to return as they would not be welcome anymore. For example, a few prisoners indicated that they need more rest, a break, and a recuperation time (Humblet and Van Limberghen 2017). In some cases, (high(er)) biological age impedes participation in prison work, e.g. due to back problems. A part of the population was thus de facto excluded from prison work. On the other hand, there are people with a high biological age (e.g. cardiovascular problems) who still exercise and work in prison (prison setting no. 1) or are able to participate in certain activities. Whilst often related, biological age should thus not be equated with functional age. Some older prisoners even cumulated two jobs in prison (such as one female detainee who worked in the prison workshop and works as a cleaner). There were also prison workers who work seven days a week ('porters'). In addition, there are older prisoners who actively work in prison, at the cost of suffering.

## Assessing Functional Age

We have encountered statements of other prison actors that touch upon the core participants' functional age. Both prison staff and fellow prisoners express their views on the functional capabilities of certain older prisoners. Against this backdrop, some prisoners are assessed as having in fact lower functional ages, which is evidenced by the following excerpts from our field notes regarding a male prisoner (Elmer, 79 years old chronologically) residing at the first prison setting:

> 'It's an attention-seeker'. I ask how? And he [prisoner-servant] answers once more: 'it is just an attention-seeker'. I ask again why and he responds: 'Just look! You see him standing there at that telephone, clinging to it.' At that very moment, I'm looking at Elmer, who appears to be on the phone and is clinging to the side of the prison phone booth with his other hand (possibly to maintain his balance?) while he seems to be browsing around. (Field Notes P1, May, 2014)

And:

> He sometimes pretends to be ignorant. [...] He also lets himself fall to get attention. (Field Notes P1, October, 2014)

Similarly, the appraisal of Anita's functional age is based on her mere involvement in prison work and other activities (Interview Anita, September, 2014):

R: Fortunately I have a lot of courage. And that I say that I want it to be clean here. But I can tell you that it's not easy. With those heavy mops, on my wrists, on my fingers. And don't think you can ask anyone else. Well, that's the sort of thing that isn't appropriate for people who are older.

I: No… And they do not help either?

R: No, they don't even ask.

I: And what if it really wouldn't be possible anymore?

R: No, they would say: 'well look, you are still able to go outside for a walk'. Because that's what the Psycho Social Services said to me. She said that it would be possible that the Sentence Implementation Court might require that I would have to volunteer a few days a week. I told the one from the Psycho Social Services: 'Volunteering? I am alone, you have to do your household, pick up your groceries, I live with pain'. Yeah, yeah, yeah. 'But you still manage to work over here'. And that was that. I said: 'Yes, because I have to'. And I also said to her: 'And now I don't have to … make sure I do my groceries and so on. Then I'm going to have to do it all on my own…'.

Several prisoners participate in prison work, programmes, or activities, despite feelings of suffering or experiencing pain. For some, this appeared to be a discursive strategy. In our study, we also encountered prisoners obscuring things that they cannot 'really' do anymore and/or they struggle with. Conversely, some tend to masquerade their functional weaknesses and are more likely to underplay the difficulties and challenges they face, which Rozario and Derienzis interpret as an assertion of their autonomy and desire to age in place (Rozario and Derienzis 2009: 551).

We [an older prisoner and I] are sitting on the edge of the bed. I am sitting right next to him during our talk. He has difficulties walking, he's slower and comes across more fragile. But he 'does his best to not let it show', he says himself. He does not want to be mutated to the medical unit, because he believes that there is no television over there. So he tries to uphold himself for better or worse, and he doesn't want to be a burden to others. He has difficulties with transport from and to prison taking him to the

> community hospital for kidney dialysis. He says that this depends from officer to officer; that some are more harsh or helpful than others. But he does indicate that 'police are even worse'; for instance in tightening the handcuffs. (Field notes P1, December, 2014)

We should therefore take into account the ways in which prisoners endeavour to influence these functional dimensions, which is inter alia also determined by their psychological age (cf. infra). Our data also suggest an interplay between the prison environment and the functionality of the prisoner in, and in interaction with, that prison environment (cf. next chapters in this regard). Conversely, other prisoners found their functional age to be lower than the functional age that was ascribed to them. It seems paramount to recognise and acknowledge all the different gradations of a person's functioning. Functional age should also be conceived as a continuum, whereby some material or social support is needed to a greater or lesser degree. The institutional thinking in prison points to the existing dichotomous thinking; i.e. it is either 'in or out'. It is therefore no surprise that some prisoners are subjected to either overburdening or underburdening.

The difficulty in establishing and assessing the functional dimension of age is also demonstrated by one of the discussions we attended in the field:

> At a certain point I observe that there is a discussion going on about the score that an older prisoner has received on a KATZ-scale [i.e. a standardised instrument to measure one's functioning and degree of dependency; this is used in determining whether a person is eligible to be admitted to a nursing home]. The prisoner stands up and hands over a letter to the prison nurse. The nurse says: 'What's that? Can I read it?'. He nods and says: 'Yes'. The prison nurse reads out loud: 'A null-profile on the KATZ-scale'. 'How is that possible?!'. 'That can't be'. Upon leaving the cell of the prisoner, the nurse says to me: 'A 80-year-old individual who has zero on the KATZ-scale, that's impossible'. The nurse heads over to another nurse to obtain redress. She then heads over to a member of the psycho-social prison service. The nurse reads that the prison doctor has written down that: 'he does need some help with toileting and showering, but that's difficult to realise in such a setting [prison]'. When reading this out loud she says: 'I wouldn't know why this would be difficult to realise in such a set-

ting?'. A discussion arose. The psycho-social service member says that he can't decide on that scale. The nurse says; 'I have to know if he can get out (to a nursing home), otherwise I won't put any more of my energy into it'. The social worker says that there is little chance he will get out. [...] (Field notes P1, July, 2014)

## Social Age

Society is made up of implicit and explicit rules regarding how one ought to behave. Social age refers to the way in which societies and cultures deal with ageing and, more specifically, how the ageing individual acts in comparison to others (Schroots and Birren 2013: 51). It covers the social roles and expectations that are embedded in society and the responsibilities or tasks that someone carries within the social structure (Hayslip and Panek 1993). Instances of a social perspective on age include dress code, lifestyle, language, behaviour, leisure activities, retirement age, and position.

In the following sections, we look at the social age of the core participants, which has been construed in terms of their familial life events, engagement in prison work, and social habits. The chronological time (cf. supra) has received content by the introduction of the concept of social time (Greve and Staudinger 2006: 807). According to the 'social clock' theorem (e.g. Neugarten and Hagestad 1976), there are explicit and implicit norms and expectations in society which dictate the age appropriateness of activities, events, and concerns, whereby this notion refers to the internalisation at the level of the individual of these social norms (Greene 2014: 125). It should be pointed out first that our data have offered very little support for explicit age-related roles within the prison society. As a result, we will mostly draw on implicit age-related roles to examine the social clocks that operate amongst older adults in prison.

### Familial Life Events

In addition to the event of retirement, for the old-age adult, the social clock of life also dictates when one should become a (grand)parent

(Antonucci and Akiyama 1997: 148). Virtually all participants have children, except for the female participant most advanced in chronological age. Most of the participants, but not all of them, mention grandchildren. It is striking that two participants from prison setting no. 1 have one or more young offspring (i.e. from preschooler to adolescent). These two men father children fairly late in life, indicating a younger social age in terms of fatherhood. Those who have lost their spouse or significant other earlier in life have a higher social age in terms of widowhood. The same holds true for prisoners' ascendants. Virtually all core participants have lost their parents. Whilst the lack of meaningful social relations often found amongst the older prisoner population could be regarded as part of the prison lifestyle for people of all ages, we question Reed's finding that widowhood is not meaningful in prison (Reed 1978: 64). Most prisoners appear to be in line with their social clock. Strikingly, there have been a few core participants who have become (yet again) parents or romantically involved with a significant other late in life, hence deducting points of their social clock. Prison might impact indirectly on the social clock of its captives by its impact on, and interference in, certain life or transition events. In this context, we would like to refer the reader to the importance of intra-muros pseudo familial relationships and/or social roles that older prisoners themselves take up, or that are attributed to them, in the prison society (see the following chapters).

## Involvement in Prison Work

Secondly, as there is no set chronological retirement age in the prison society, prisoners' social clock in terms of their association with work is either advanced or delayed by other types of age in synergy with the prison environment. It should be noted that older prisoners are not considered to be entitled to adapted employment within the realm of the prison: the inhabitants of the second prison setting were de facto and excluded from engaging in prison work by the prison system on grounds of an appraisal of their functionality which is a priori implied by these prisoners' admission to the gero-medical unit, whereas older prisoners in the first prison setting strongly had to adhere to the norms that have been geared towards younger workers, although not a priori excluded. However, we found that older individuals in the first setting relatively often solicited the exceptional status of prison-servant/porter.

Her scrutiny of the concept of 'retirement' in a prison setting in the United States[6] had already led Reed (1978: 13f.; 44–49; 61f.) to argue that retirement has little meaning within suchlike setting, as there is no set age restriction or limit on working. To a certain degree, the meaninglessness of the concept of retirement in prison has also been borne out by our study, as there is no set retirement age for prisoners in Belgium. Although prisoners are not obliged to work during their incarceration under the Belgian system, it must be noted that they do not accrue pension rights during their detention, nor do they receive a pension payment whilst imprisoned despite their general eligibility (Humblet and Van Limberghen 2017). Rather, the opportunity to bestow labour in prison is fixed by a complex interplay of bio-medical (functionality, biologically fit for work), economical (turnover, supply, and demand), and/or prison constructs (e.g. discipline, safety).

As for the concept of retirement[7] in the integrated setting (P1), we note that approximately half of the male participants and all three female participants still remained involved in some kind of paid work in the prison setting whilst exceeding the general retirement age in free society. Conversely, as for the segregated setting (P2), our study has revealed that all participants in the segregated prison unit are de facto retired; that is, they are excluded from performing any form of (paid) labour in the prison setting, despite having a similar chronological age as that of the population housed in the integrated setting. For those older prisoners who had been involuntarily excluded, work also still played an important role, albeit rather in the form of the past (to retain their identity). When using retirement as an indicator of old age, in this respect, the second group (including de facto retired persons from the first group) could be described as having higher social ages. For some, this was prompted by their functional incapacity to work. In both examined settings, there have been both prisoners expressing an intention to resume their previous (employment) activities upon release (irrespective of any realistic self-appraisal as to the feasibility of their aspirations) and prisoners who did not wish to do so. The second category of prisoners is excluded from prison work but were supposed to take part in activities that are designed for them. This brings us closer to the following passage, wherein it is examined how the core participants construct their social age in terms of social habits, to which activities pertain.

## Social Habits

The social age of persons refers, inter alia, to their activities, attitudes, preferences, patterns of dress, speech, and deference to persons in leadership positions relative to the expectations of society (Schroots and Birren 2013; Hayslip and Panek 1993: 9). Social age in terms of social habits tends to have a profound impact on older prisoners' relationship with peers and/or younger counterparts in terms of in- or exclusion in social prisoner groups. The second setting was comprised of higher social age norm groups in contrast to the first research setting. In both settings, some older prisoners denounced the type of activities offered to them. It seems to follow that (older) prisoners with analogous social ages are generally less discordant.

In addition, certain participants have touched upon their implicit assumptions on how to behave socially ('according to their age'). One female participant perceived herself to be in step with the social clock but clearly perceived others to be out of step with their age ('she does not act her age'), as evidenced by the following accounts:

(Interview Anita, September, 2014):

I: So you were saying that you also have a 'moral punishment'?

R: Yes, by the fact that I... You obviously have a moral punishment because you are so much older. Yes. And you have the punishment. Because you are older, because I feel that I do not fit in. It's just that way. It's not just claptrap. I am not the type that complains. No. So, 'tis like that. You have an ordeal, by the fact that you are so much older, and also your punishment, huh. But the fact that you are not coming into one's own... I would also like to talk occasionally. I also want to go for a walk. You get laughed at.

I: You mean by the younger ones, or?

R: Yes. Yes, yes, yes, yes

I: And what about the ones who have reached a higher age, like... Dorothy and?

R: She doesn't do that.

I: Do you sometimes talk to her?

R: Hmm, no ... I talk to her, but ehm... She's totally different. She adheres to the... To? To the youngsters... Well, I don't know if that's innate.

But... I, I feel young at a certain time, in a certain way. But not participating in it like 'gihihi, haha'. I will also laugh if necessary, but ehm, yeah I'm different. Yes. I'm different, I'm ... Well, for my age, I am different. Yes...

(Interview Anita, November, 2014):

R: Take Dorothy, yes, she is older, but 'tis also a child. She keeps childish with the children. If she feels good about that..., agreed, but I doubt whether it is of ... Well, to fit in. I reckon it must be something like that. I cannot do that. I am 70, but I won't...

She also does that, talk about that sex-stuff and about that... I will not do that, huh. Of course, she is ... well yes, she fits in then.

I: Yes, I understand.

R: Perhaps it can be beneficial to her, that might be true. But, no... I will not chat all day long about those men and about that sex and drugs and about the murders they committed.

One of the older female prisoners epitomises the paradox of growing older but staying younger, facial features of an older person, but dresses, acts, and speaks similar to younger prisoners (e.g. also importance of dyed hair for some, whereas one male prisoner stopped dying his hair in prison, therefore increasing his age as perceived by others in terms of appearance). These apparent outwards features seem to play a more prominent role in the conceptions of old age in relation to the older female prisoners included in our sample.

One's social age is a characteristic of the participants that also affects their connections with other prisoners. A generational gap might be overcome when a prisoner can form a bond with another prisoner based on a shared commonality, such as being 'ostracised' or having a 'minority status' within prisoner.

Bearing in mind a chronological age differential of 13 years and a differential in cognitive and biological functioning, Ivan (male, 69 yrs) expresses during an interview why he is rather disturbed by the behaviour, attitude, and clothing preferences of one of his 'age peers' (Virgil, male, 82 yrs) who is incarcerated in the same prison unit and is denoted as old (Interview Ivan, November, 2014):

> For instance, you have one case over here, Virgil. Do you know him? The one who speaks so loud. Well, talking with him, I wouldn't even recommend that. He has his own ideas. As far as he's concerned, the woman's place is at the fireplace and the man has to work. Women who work, they are nothing worth.
>
> […]
>
> And he always wears short trousers, huh. That was his hobbyhorse. He said: 'I wonder if I can still wear my short trousers at the nursing home'. […] You have people over here that wear short trousers, huh, and that's always ridiculous. Because ehm… I've known older people who wear short trousers, it's unbelievable.

Others with similar chronological ages showed diverging interests. In this regard, we have noted that some core participants played video games and demonstrated technical interests and skills (such as computer skills), with which older people are not often associated. These prisoners maintained lower social ages. Conversely, there were chronically younger prisoners who were constructed in terms of higher social ages (e.g. playing card or board games). The inter-individual variability amongst prisoners in terms of their social age at times poses issues in both settings if one's social age is aberrant vis-à-vis those of others and it cannot be compensated for by one's emotional age.

Biological ageing can have an effect on social age. A decline in biological functions such as sensory impairments (e.g. hearing and/or visual impairment) can impose restrictions upon older people's ability to engage in interactions with others (Jiang et al. 2016: 37). This has been epitomised by Anita (Interview, September, 2014):

> The worst is when you're among a group of prisoners, and during airtime. Well, see, it's the same again. If I go to something … I'm not going to the crea, because I cannot hear them. Because they chat in a group, so I can't hear it. You see? Then they'll say: 'You are not going to the crea?'. No,

> because I cannot hear it! If there is something, I won't go because I cannot hear anything [she is getting a little stirred up]. Because it takes place in group and if they speak in group, I can hear them producing sound, but... See, these are all things that bring about loneliness. [At this point the conversation gets emotionally charged].

Our data suggest that there are both too little expectations towards older prisoners, i.e. that they cannot learn anything and that they aren't interesting ('their time is worth less'), whereas prisons might also have too much or excessive expectations, e.g. finding employment to re-integrate. Moreover, there were also expectations from older prisoners themselves, phrased as 'I am too old to take this crap' or by using similar expressions (Interview Dorothy, October, 2014), whereby some prisoners tend to appropriate to themselves what could be subsumed under the denominator of 'seniority'. This refers to something that accompanies advanced chronological age and prescribed how to deal with old-age adults.

In prison, there are implicit expectations related to older prisoners' involvement in deviant behaviour: such as that they are compliant, pose no problems in terms of security and order, and have little future prospects. Moreover, the idea that 'older prisoners should have known better as a result of their old age' has also been expressed (Interview Medical Staff Member, December, 2014). This particularly has an effect on those who have been (re-)admitted to prison at a chronologically advanced age.

## Psychological Age

Psychological age comprises behavioural and perceptual aspects. Behaviourally, psychological age is based on the adaptive capacities and skills that people have and use to adapt to changing biological and environmental demands (such as memory, intelligence, problem-solving, coping, emotions) (Hayslip and Panek 1993). Moreover, psychological age is comprised of cognitive age, i.e. one's abilities to learn and remember, and emotional age, defined as 'one's ability to handle and manage feelings' (Whitbourne 2012). More subjective dimensions of ageing are also subsumed under the perceptual component of psychological age, which refers to how old people perceive or feel that they are (e.g. 'subjective age', 'feel-age') or how old they are perceived by others ('others-perceived age') (Hutchison 2014).

## Behavioural Aspects

(Older) prisoners can suffer from cognitive declines that, for some of them, herald the onset of dementia. The finding that high cognitive age is not the prerogative of 'old age' has been clearly demonstrated by our encounters with prisoners who had not reached the chronological age of 65 years yet were constructed in terms of cognitive impairments, even dementia. At the time of the data collection, such older prisoners seem evenly distributed amongst both settings; approximately the same number of core participants with signs of dementia has been discovered in both settings (three and two in the first and second prison setting, respectively). However, this number was not proportional to the total number of older prisoners housed in each setting. The phenomenon of dementia has been more distinctly present and displays itself more prominently, in the 'gero-medical' prison unit. This study supports evidence for the many difficulties that exist in prison as to diagnosing and treating dementia, due to, inter alia, limited availability or capacity of resources and/or staff, staff lacking the training to recognise early warning signs of dementia as opposed to normal ageing, and/or the regimented structure of prison life that can mask (early) signs (Moll 2013). In addition, some of the core participants who had been diagnosed with dementia also tended to 'mask' their disease or some form of deterioration herein. During our fieldwork, it became evident that similar diseases or ailments might manifest itself differently amongst individuals, which has been pointed out explicitly by one of the prisoner-servants residing at P2:

> (Interview Peter, March, 2016):
>
> Those are the four people that have it … albeit each in a different way. It [dementia] does not present itself in the same way, but those are the people that have like, ehm, problems with cognition, with memory, and living in the past. But the reason behind it, and how it manifests itself, is different

> for each of these persons. [...] But ehm ... those people crave for a clear structure. And you notice that in every way.

The second prison setting has appointed a nurse who tries to cater for these prisoners. For instance, specialised games are played with some of the older prisoners residing at the unit, which are designed to benefit those with memory problems (Moll 2013: 23). A major concern is to what extent prisons fully acknowledge or address this phenomenon. Individuals with (alleged) signs of dementia have been found in both prisons. This might also raise questions as to the alleged benefits of specialised health care, as prisoners with all kinds of conditions and needs are housed together, even within the segregated setting.

The second strand of the 'psychological age' of the core participants encompasses their emotional functioning. Most older prisoners involved in our study tended to have reached higher emotional ages. Those with higher emotional ages are deemed to have gained some higher level of 'psychological maturity': i.e. they are better positioned to regulate their emotions (interaction with themselves), e.g. control their temper; take (a) negative situation(s) and spin them in a positive light (interaction with situations); get along more easily with other people even with those they don't particularly like (interaction with others); and distinguish between their own thoughts on the one hand, or in other words their inner world, and the outside world on the other hand ('distancing'; interaction between self and the environment) (Whitbourne 2012). It should be noted that attaining higher levels of emotional age is not reserved for, or limited to, older prisoners, but there can be a higher likelihood to attain it in old age. This tends to be beneficial in a prison context, which evokes emotional responses. Having a high(er) emotional age facilitates coping with the ageing-imprisonment nexus, which will be discussed in one of the following chapters.

Amongst others, Ivan's account bears witness to his (Interview Ivan, November, 2014):

> You have to act how they want you to. You cannot act excessively, but, well, that doesn't interest me at my age. [...].

For some, biological age might be therefore considered secondary, or susceptible, to this psychological dimension:

> (Interview Antonius, March, 2016):
> I: Do you feel healthy?
> R: Yes, I feel healthy, yes. I cannot complain. I have to accept it. I do already have a certain age … 79 in fact! I do not have to start dwelling on getting ill and sickness and health, huh … I… No.

Information on the behavioural aspects of ageing in relation to the experience of imprisonment was analysed on the basis of four types that have been identified by Baumeister and Keller (2011: 52–72) in their research[8] on older prisoners in Swiss prisons (see also von Borstel et al. 2014), as outlined below:

1. *The trying preserver* ('Der Bemühte Bewahrer'): this prisoner type identifies himself with the status he had in his former life and tries to uphold it. This type of older prisoner is moreover characterised by high social skills and is respected and accepted within the prison. It is also significant that his relationship with staff is largely devoid from conflict. He distances himself from the prisoner community, with prisoners seeking contact with him and not vice versa. He maintains external relations outside the prison walls. He had been admitted to prison for the first time in later life, has a punishment with a foreseeable end, and can look back on half a year to six years of imprisonment (Baumeister and Keller 2011: 52–58).
2. *The conformist restarter* ('Der Angepasste Neubeginner'): in contrast to the first type, this prisoner is ashamed of his former life and breaks off ties with the outside world. He moreover actively participates in the treatment programmes offered by the institution and tries to build new relationships with other inmates. Representatives of this group view imprisonment as a positive turning point in their lives and have built up new contacts from prison, with whom they want to reshape the foreseeable future. However, this strong optimism can also be accompanied to a certain extent by scepticism as to whether the 'new beginning' after the release from prison actually will succeed on the

outside. Whilst their response to imprisonment is reminiscent of what Goffman denoted as 'colonisation', Baumeister and Keller stress the difference that this type of inmate places the newly acquired future after the detention at the centre. This second category is comprised of a group of repeat offenders and long-term prisoners with a total of 10 up to 30 years of prison experience (Baumeister and Keller 2011: 58–62).

3. *The misunderstood outsider* ('Der Missverstandene Ausgeschlossene'): this type of prisoner has known little social acceptance by others throughout his life. As such, he lived rather withdrawn and often has had an affinity with subcultural communities. The mere act of imprisonment consolidates this social exclusion. He feels misunderstood by laws and by society. This type is also characterised by few outside relationships, few contacts within the prison and is mostly on hostile terms with the prison staff and other inmates. He spends most of his time alone in his cell, with biographical and justifying written accounts or other self-employed activities. In addition, he tends to actively justify the crime he was incarcerated for. These inmates are mostly uncertain as to their end of sentence; and only a few have been imprisoned for the first time at an advanced age (Baumeister and Keller 2011: 63–67);
4. *The unnoticeable lackadaisical* ('Der Unauffällig Resignierte'): this type of prisoner has been subjected to multiple problems for over an extended period of time. Continuing in the same vein, this older prisoner has few personal, social, or financial resources to change his current prison plight. He barely has existing social contacts (both intra- and extra-muros) and is often denoted as a 'loner' ('Einzelgänger'). This prisoner tends to be resigned, has little future perspective, and views prison as a further station, with which he has to come to terms with. He acts very inconspicuously and is oftentimes overseen by other inmates and the prison staff. Prisoners belonging to this category are either recidivists who have received fairly shorter sentences or inmates with one single offence who have already spent a fairly great deal of time in prison (Baumeister and Keller 2011: 68–72).

In our own research, we have identified these types of older prisoners amongst the prisoners that have been included in our sample, with the former and latter type being the most notably represented. Some types have however been less prominently present, or more diluted, amongst the core participants of the study, thus preventing us from drawing any well-established conclusions as to this type of older prisoner (respectively, the second type as outlined above). In terms of emotional age, we have established that participants who could be subsumed under the former and latter categories (respectively, categories one and four) generally have shown higher emotional maturity in dealing with age-related processes and/or imprisonment. As to the link with *incarceration history* and *age at admission*, we first note that recidivists and first-time prisoners are dispersed over different types and have different ways of dealing with imprisonment and experience differences in the way they are being dealt with. In addition to this, it has come to light that life before being admitted to prison is also of value (i.e. in terms of different ages and the individual life story), rather than chronological age in itself.

## Perceptual Components

With regard to late-life identity, the idea of the 'ageless self' has been established, which refers to the proposition that later life has no intrinsic meaning, and older people tend to see and present themselves rather as the continuation of younger identities than as 'old' (Rozario and Derienzis 2009: 541; Biggs 2005: 120). In this view, older people's dilemma is that of a youthful self 'trapped in an ageing body' (Biggs 2005: 120f.). In the same vein, the notion of masquerade has also been suggested, which is deployed when an older person acts out the role of the younger person to protect a mature identity (Biggs 2005: 121). However, both these views have been contested for neglecting the influences of ageism and stigma in the efforts of older adults to avoid being labelled old (Andersson 2002: 72; Rozario and Derienzis 2009: 541). The view of Sugarman captures the essence of the argument here, that by acknowledging and disclosing the significance of age, it enables us to look beyond the *mask of age* to the *agelessness* as well as the *ageing* of the person beneath (Sugarman 2004:

107, own emphasis). Physical ageing, as an identifier of old age, has been referred to as the 'mask' of old age, a sign to others as well as oneself, behind which one's inner 'true' identity lies (Pain 1999: 3). It illustrates how the concept of age has *objective* and *subjective* components. Objective age relates to the measurable or observable qualities, whereas subjective age tends to concern itself with experience, perception, and identity. In addition to the objective and/or more traditional considerations of age (cf. supra), a more phenomenological perspective on ageing draws on the subjective experience of age (Settersten and Mayer 1997: 242). People thus have their own frame of reference based upon how they experience the ageing process.

## Subjective Ageing: Age Identity

How individuals experience 'ageing' has become a significant strand of scholarship and is related to the idea of empowerment. In studies of the subjective experience of age(ing), the concept of age identity has emerged and refers to the inner experience of a person's age. It is regarded as 'the outcome of the processes through which one identifies with or distances oneself from different aspects of the ageing process' (Westerhof 2008: 10). Individuals may identify themselves with certain age or age groups (*identity age*), prompted by, for example, how old they feel (*feel-age*). This provides a further nuance to ways of looking at age (Settersten and Mayer 2002: 33).

A large body of empirical work has highlighted the tendency of 'older' individuals feeling younger than their chronological ages (see, for instance, Chasteen and Cary 2015: 100; Westerhof 2008: 11). A study of the discrepancy between chronological age and felt age amongst community-dwelling older adults indicated that older adults, who face declining physical/functional health, have more reasons, and greater motivation, for feeling younger than their chronological age than younger adults (Choi et al. 2014: 468). Whilst disabling conditions were found to be a more significant correlation to of chronological and felt age discrepancy amongst the middle age group of older adults (i.e. for the 70 to 79 age group), this appeared not to be the case for the youngest and oldest group (i.e. 65 to 69 and the 80+ age group) (Choi et al. 2014: 469f.). As for the former, the young olds, it is surmised that they must suffer from fewer, and/or yet less severe, disabling conditions, or they might have chosen to focus on other aspects of life than their disabilities. As for the latter, the 80-and-older group, it is suggested that they may make selective and strategic comparisons to focus on aspects of life that transcend their physical/functional abilities (Choi et al. 2014: 469f.). According to these authors, people tend to engage, both intentionally and unintentionally, in a downward comparison, so as to enhance their own subjective well-being by comparing themselves with less fortunate others (Choi et al. 2014: 469). It is also common amongst older adults, both in Western and Eastern countries, to feel younger than their chronological age, a tendency that is more pronounced with increasing age (Choi et al. 2014: 459). Similarly, Rozario and Derienzis (2009: 550), in their examination of how age identities are constructed amongst older people (65+) facing chronic conditions, found that regardless of the commonality of chronic conditions, participants associated different meanings to that experience and their understanding of age, which enabled them to develop different age identities. They pointed out that, notwithstanding their experience of chronic illness, the majority of respondents continued to actively present themselves as 'not old' (Rozario and Derienzis 2009: 549). Consistent with these findings, Prevc and Doupona Topic (2009: 1110) also found the subjective age of nursing home residents to be significantly lower than their chronological age.

This section on the core participants' 'age identity' deals with *their inner experience of their age and ageing process* (Westerhof 2008: 10), which might have been elsewhere subsumed under a myriad of terms, such as feel-age, self-perceived age, and subjective age. Our results corroborate the findings of a great deal of the existing body of research on old-age persons' experiences of ageing (Westerhof 2008: 10f.), as the older prisoners involved in our research also tended to feel younger than their chronological age. There are similarities between the attitudes expressed by the prisoners in our study and those described by Reed (1978), which showed that an overwhelming majority of older male prisoners (i.e. denoted as having a chronological age of ≥50) in her sample felt 'young to middle-aged'. The prisoners in her sample even felt younger compared to people on the outside (Reed 1978). The occurrence of temporal, somatic, social, or mental phenomena that bring some of the ageing processes into sharper focus does not preclude core participants' subjective experience of feeling younger than their calendar age. Strikingly, the subjective age identity of most sampled older prisoners was found to be *significantly younger* than their objective chronological age. The following transcript excerpt, taken from an interview with one of the—calendar wise—oldest prisoners housed in the gero-medical prison unit, even provides one of the more extreme illustrations hereof (Interview Albert, February, 2016):

I: How old are you?
R: Seventy- … Seventy-three [whispers].
I: Really?
R: Yes. Within a couple of months, I turn seventy-four.
I: And how old do you actually feel?
R: Twenty-five.
I: Yes?
R: [Laughs]. Yes, I'm still … I still feel good.

More representative instances of older prisoners distancing themselves towards different aspects of the ageing process are epitomised by the following answers to the respective questions posed: i.e. 'How old do you feel?' and 'What age do you feel you are?', which are considered particularly valid questions to probe into person's subjective experience of ageing (Westerhof 2008: 10):

> Me? Ghosh! I'm too young to say … I think I'm about 45 years old, or 40 years. Yes … And then I sometimes think about my time as a soldier. (Interview Antonius, March, 2016)

And:

> Well, I'm not sure about that. I'm 70 years, but whether I feel 70? … I can't compare. (Interview Ron, March, 2016)

Such discrepancies might be understood as the result of *adaption* to *age-related* changes, whereby two processes are distinguished: that of *assimilation* and that of *accommodation* (Westerhof 2008: 12). In the former individuals strive to maintain their existing identity in the wake of (age-related) changes, whereas in the latter, individuals may react to such experiences by changing their identity (Westerhof 2008: 12). The phenomenon of adults feeling younger than their actual age is from this perspective regarded as the result of an assimilation process wherein they maintain an existing identity (Westerhof 2008: 12). It could also be argued that the prison environment mirrors a wider social context in which youth is a more valued status than old age, as a result of which old-age prisoners identifying with younger ages might even endeavour to achieve some self-enhancement next to attaining a sense of self-continuity over time (Westerhof 2008: 12). We will return to this in one of the later chapters.

Consistent with the literature on self-identity in late life (Matthews (1979) in Rozario and Derienzis 2009: 541), we also found that some of the participants, who rejected the label 'old' for themselves, simultaneously used it on some of their age peers. An 'older prisoner' who strongly expressed his unwillingness to accept the label of 'old' for himself frequently tapped into this vein to distance himself from his age-peers or distinct processes associated with ageing: e.g. 'Oh give me a break, can you picture me amongst someone as "Christopher", the one who drives with its trolley' or 'the old notary' or 'that old lad who works in the kitchen' (Miscellaneous Field notes). Our data moreover tend to support the difference between one's *social* and *personal* identity, of which the former refers to 'how people see and define themselves in relation to their membership in a social group or category', whereas the latter includes 'the idiosyncratic and unique attributes that distinguish the person from others' (Rozario and Derienzis 2009: 540). In this respect, we observed that many core participants did not have the feeling of belonging to an older age group and that they were more prone to see themselves as an exception, similar to older adults in general (Westerhof 2008: 10f.). An older prisoner who distanced himself from others and from ageing as a process of decline references to himself one of the prison officers: 'you see, this old-timer is still briskly' (Field notes P1, April, 2014). Whilst some older prisoners have expressed to feel much younger, they do speak at the same time in terms of their alignment with a social 'older age group' as well. For instance, a 79-year-old prisoner with a feel age of 40 years nevertheless mentioned that 'we elderly people are calmer' (Interview Antonius, March, 2016).

With regard to aged prisoners' 'self-concept', Reed found that prison places a great deal of emphasis on a person's physical and mental condition in which case and how prisoners are treated as an 'older' prisoner, the men in her sample reported by the same token that certain events give rise to the feeling of being older, e.g. becoming grandfather or as a result of injury (Reed 1978: 49ff.). Reed's previous finding that chronological age does not appear to be a major factor in a person's age definition (Reed 1978: 50) resonates with the results of our research.

One of the male respondents alluded to the notion of a stable identity:

> I: And do you still feel young?
>
> R: As a matter of age? I think that that is something… If you take the physical wear out of consideration, I think that's something that does not exist; 'age'. Apart from the experience that you might also have of course. But, in my opinion, that is something that not exists. If you were to live in the same young body… I don't know, apart from that little bit of wisdom you gain with the years… And do they call that old age, then? I don't know. What do they mean by that, old age? Don't *you* know it? (Interview Gerald, November, 2014)

The way in which this prisoner defines his subjective age bears a significant resemblance to the phenomenon that is known in the gerontological discourse[9] as the 'ageless self' by Kaufman (1986), or as 'the mask of ageing' (Featherstone and Hepworth 1991) pointing at older individuals reporting strong stability in the sense of a 'Self' that is unchangeable in spite of bodily changes. According to Featherstone and Hepworth (1991), the visible body 'masks' the unseen 'youthful' Self. Conversely, the 'masquerade of ageing' (Biggs 1997; Woodward 1991) points to the protection of a mature identity in the context of an increasingly ambiguous external environment, whereby youthfulness is 'giving the right to be seen and heard' (Woodward 1991: 149). Both views have been somewhat addressed in this chapter (cf. supra on the opposing views of two chronologically aged female prisoners on which one of them is 'real' in old age and true to her inner 'Self').

Another interviewee makes reference to his own perspective on *age(ing)* in one of the interviews, by stating the following (Interview Antonius, March, 2016):

> I: Do you notice that you're getting older?
> R: Yeah, as a result of my birthdays, but otherwise … pfft, no…

It has been already pointed out that instrumentally measured time (via chronological age) might give rise to social comparison amongst individuals (Baars 2012: 42f.). The social comparison theory is premised on the idea that people use others as a reference point to understand and evaluate their own abilities, opinions, and performances (Carmichael et al. 2007: 292; Festinger 1954). We note that prisoners also measure themselves with others to assess themselves. We will return to this in one of the later chapters, wherein social comparison is employed as a coping strategy amongst older prisoners.

## Others-Perceived Age

Age as perceived by others gives also a subjective meaning to age, the only difference being that the evaluation is done by others. It is concerned with the subjective evaluation of the age status of an individual by others (Barak and Schiffman 1981: 602ff.). According to Barak and Schiffman (1981: 602ff.), this type of age would be especially appropriate when examining stereotyping of age groupings and is largely based on perceived physical looks and social roles of individual(s). Analogous to the reasoning of Barak and Schiffman (1981: 602ff.), this type of perceived age is especially apt for exploring if, and how, prison actors' perception(s) of a prisoner's age interacts with relational aspects and other relevant environmental or situational factors.

During our ethnographic fieldwork, prison staff have often been mistaken in designating or omitting certain prisoners as 'older', albeit that 'old age' in itself has also been discussed amongst prison staff. Several prisoners with advanced chronological ages (i.e. 65 years and above) have not been identified by other prison actors as such, whilst chronologically younger prisoners have been denoted as 'at least 65 years old'. Prison staff members have been found to draw on physical features and appearance to visually estimate the age of a prisoner.

When mistaken a chronologically younger prisoner for an 'older' one, it has been often ascribed to 'prison harm':

> In the process of identifying 'old age prisoners', they [i.e. prison officers] are often mistaken. This becomes evident when they look up the date birth of individual prisoners that were identified as 65-plus or over. The prisoners that have been identified appear to be 50+ or approximately 60 years old. [...] A male prison officer notes in this regard: 'Yeah, you know, that's due to prison harm. Some [prisoners] look a little decrepit.' (Field Notes P1, August 2014)

It is an analogous observation of several correctional officials 'that the typical inmate in his fifties has a physical appearance of at least ten years older' (Aday 1994) which has precisely given impetus to the scientific community to examine the idea of 'accelerated ageing' amongst prisoners.

Conversely, it has been other researchers' contention that older men in prison even look younger as compared to older people in the community (Reed 1978: 63f.). Our examination of the perceived age has shown that several of the older prisoners involved in our study have not been nominated as being of (highly) advanced age. Despite him suffering from a chronic disabling disease of the central nervous system, and being visibly bounded to a wheelchair, Albert was one of the participants who took particular satisfaction in others (under)estimating his (chronological) age. Like some others, Albert expressed to exhibit no signs of difference in the progression of ageing as a result of his incarceration:

> (Interview Albert, February 2016):
>
> I: Do you think that being imprisoned has had an impact on your health, or on the process of ageing?
>
> R: No ... no, not at all. Because when I say how old I am, they are startled [laughs].

Moreover, we found that this perceived age might also refer to other (non-)chronological dimensions of age. From older prisoners' point of view, a significant discrepancy between others-perceived age and other dimensions of age is both experienced in terms of advantages and drawbacks. Inmates' (un)fruitful attempts to manage some of the physical characteristics in the wake of the ageing process have been determined by prison life (e.g. male prisoner who has stopped dying his grey hair since incarceration) (Field notes P1, December 2014). For many older prisoners, the others-perceived age was generally found to be lower than their chronological age. These prisoners might be to a lesser degree prone to encounter the detrimental effects that are associated with an 'old-age status' in prison. Conversely, low(er) others-perceived age(s) may mask prisoners' higher different age dimensions, which will be discussed further in this chapter. In such cases, the vulnerability that accompanies old age—albeit to different extents—is left unaddressed and is therefore exacerbated. In support of the latter result, one female prisoner comments on some of the adverse effects of her lower(ed) others-perceived age, thereby disregarding her real struggle:

> And of course, they don't see that; because you continue to work here, but when you complain they'll say, 'She's still good for her age, huh?', but you're living with pain and so on… That's just the way it is. That superficiality…. (Interview Anita, September 2014)

## The Older Adult in Prison

Political philosophers might refer to this book as 'widening the circle of moral considerability'—which is one way of thinking about the prison experience for older adults. Asking who is chosen as a subject makes for deeper, more fundamental claims.

In 2011, Liebling and Maruna wrote in their introduction to *The Effects of Imprisonment*:

> Prison is not a uniform experience. Studies have tended to take undifferentiated samples and to look for general patterns. These general studies

> neglect the experience of particular groups and individuals, such as women, the young, the old, prisoners segregated for their own protection, those spending long periods of time in segregation units for other reasons, and so on […]. (Liebling and Maruna 2011: 12)

The pains of imprisonment experienced by older adults have only recently received limited penological scrutiny. Drawing on the deprivation and mortification model, the existing body of research suggests that older prisoners represent another distinct population which faces a new range of 'pains' that both consolidate and exacerbate those experienced by the (general) prisoner population. International—predominantly Anglo-Saxon—studies suggest that older prisoners provide an example of a group who experience 'additional pains of imprisonment' (Mann 2012) and is subjected to 'hidden injuries' that are generated by the prison regime and environment (Crawley and Sparks 2005). These hidden injuries are described by Crawley and Sparks (Crawley and Sparks 2005: 351) as 'un-noticed injuries to the self that arise in the course of everyday institutional practice, including in prisoners' interactions with staff'. They discuss how the fact of imprisonment intrinsically weighs differentially on the older prisoner—the fear of dying in prison, or of being released into insecurity and isolation, the sense that prisoners report of being irrevocably cut off from the past. Similarly, Mann (2012) stated that the generic pains of imprisonment for ageing individuals are intensified by being incarcerated in an environment which is both deeply masculine and designed for aggressive young men. She argues that older men thus do 'harder time' and experience imprisonment as a more severe form of punishment than their younger counterparts.

Whereas some hidden injuries can be related to the distressing effects of imprisonment itself (e.g. entry shock; fear of deterioration (illness); death and dying; separation from significant others) (Crawley and Sparks 2005; Glover 2012: 162), others are caused by what will be termed as 'institutional thoughtlessness' (e.g. attitudinal barriers to full participation in prison life and feelings of powerlessness and invisibility) (Glover 2012: 162). These additional or specific pains of imprisonment are created because the prison as an institution is not designed for older prisoners, who bring age-specific characteristics which prisons do not cater and who suffer from additional pains. Although Mann herself did not use the deprivation and importation models to interpret this idea of 'additional pains of imprisonment', it could be argued, for example, that the importation model could be employed to explain how the age-specific characteristics, which ageing prisoners bring with them and are thus imported into the prison setting, create extra problems.

Institutional thoughtlessness is defined by Crawley (2005: 350) as "the ways in which prison regimes[10] (routines, rules, timetables, etcetera) simply roll on with little reference to the needs and sensibilities of the old". Although this term was first introduced in 2005 (Crawley 2005; Crawley and Sparks 2005), similar instances of inadvertence or indifference vis-à-vis older adult prisoners have already been touched upon. For instance, Golden (1984) drew upon her experience in a medium-sized jail in discussing the problems encountered by older inmates as well as personnel. The problems she described at the time bear witness to the fact that jails are designed for a much younger population. She points out issues regarding medical screening upon admission, the physical layout, programmes that are geared to younger inmates, adjustment to daily life in jail, and interpretations of staff (of signs) (Golden 1984: 145–147). Similarly, Goetting (1984b) also denounced the lack of special policies, programmes, and facilities for older inmates in the United States. She found that prisons do not commonly recognise chronological age status independent of *health status* and that programmes, facilities, and treatment based on age are rare and never comprehensive (cf. previous section on reflections on these types of age). Additionally, she tentatively suggests that this reflects a lack of recognition of, and responsiveness to, the special needs of institutionalised senior citizens. The prison structure is designed to accommodate the general age category (Goetting 1984b: 171–172): "The mainstays of prison programming are education, vocational training, and recreating, and the programs offered are typically of little value to the elderly inmate" (Goetting 1984b: 172).

## Institutional Ageism in Prison

It has been suggested that, beyond the notion of physical prisons, we create and maintain our own 'prisons' by virtue of social phenomena, including ageism (O'Brien 2011: 522). Whilst negative attitudes towards old age and/or ageing had already been identified much earlier (cf. Achenbaum 2015), it was not until the late 1960s—notably when Butler[11] first coined the term 'Age-Ism' to denote age stereotyping and discrimination against old-age adults—that this phenomenon gained meaning as an affliction (Achenbaum 2015):

> Ageism can be seen as a systematic stereotyping of and discrimination against people because they are old, just as racism and sexism accomplish this with skin color and gender… Old people are categorized as senile, rigid in thought and manner, old-fashioned in morality and skills… Ageism allows the younger generation to see older people as different from themselves; thus they subtly cease to identify with their elders as human beings. (As cited in Butler 1989: 139)

In the wake of Becker's *Denial of Death* (1973), it is posited that prejudice against older adults are employed as a primary defence mechanism against existential anxiety, as they could be regarded as the ultimate symbolic representation of upcoming death (Youdin 2014: 67). However, the literature on the theoretical underpinnings of ageism has identified both the fear of *death* and of advanced *ageing* as primary sources of in-group bias and out-group prejudice (Youdin 2014: 67):

> Ageism existed long before it was given its name. Fear of and distaste for growing old are inherent in the human experience. They are not fully accounted for by fear of death. With aging comes the prospect of a decline in functions such as memory, sexuality, and mobility, as well as the occurrence of age-related diseases. The fear of deterioration, dementia, and dependency is powerful. (ILC 2006: 17, own emphasis)

Whilst Butler indicated that ageism could work in both ways, i.e. older people could be negative towards youth in the same way as younger people towards the old, the former constitutes the most prevalent examples (Butler 2005: 85). In addition to the established negative meaning of ageism, there has been some recognition of positive ageism, i.e. prejudice and discrimination in favour of the aged, albeit less common (Palmore 1999: 6; 34).[12] Butler views ageism as:

> … manifested in a wide range of phenomena, on both individual and institutional levels—stereotypes and myths, outright disdain and dislike, simple subtle avoidance of contact, and discriminatory practices in housing, employment, pension arrangements, and services of all kinds; and epithets, cartoons and jokes. (Butler 2008: 25)

Beyond the differentiation between institutional and personal ageism, the literature distinguishes between intentional and unintentional (or (in)advertent) forms of ageism (Youdin 2014: 67). Some argue that the prison system is actively and *overtly* ageist in relation to older *female* prisoners (Wahidin 2000: 55, 2005: 11). However, what has been described in much of the (prison) literature is strongly reminiscent of unintentional institutional[13] negative ageism, i.e. it is a consequence of *rules and practices* embedded in the prison institution *without recognition* that are discriminatory *against* older people.

By drawing on the work of Clemmer (1966), Goetting (1985) probed the parallel between prison society and the free community in America in terms of three distinct minority relations: racial, gender, and age. Her theoretical analysis, which was grounded in formal and informal prison practices, led her to suggest that ageism in American society was reflected in the prison structure by neglecting the particular needs and interests of the aged population:

> While the inmate peer network may assign prestige to its elderly members, these older prisoners hold no such position of esteem in their relationship to the formal institutional structure. In the prison setting, the problem for the elderly is not so much one of differential treatment, however, as it is one of negligence. […] Negligence toward the elderly becomes evident with the nature of recreational and educational/vocational programs which clearly cater to the predominantly young inmate population. They typi-

> cally are not conducive to participation by elderly inmates. Recreational programs (for men) emphasize strenuous and competitive sports, and educational and vocational training emphasize future employment. (Goetting 1985: 18, own emphasis)

She furthermore concluded that 'neglect', which she ascribed to apathy rather than hatred (as against blacks), or paternalism (as against women) persists as the most common form of ageism (Goetting 1985: 19). In addition to the negligence of older prisoners (cf. ageism as unintentional neglect/unjustified equal treatment), she hinted at signs of outright discrimination (cf. ageism as discrimination/unjustified unequal treatment) in terms of employment opportunities and pay (Goetting 1985: 19).

The literature attributed these pains to a lack of regime differentiation and unsuitable prison environment (cf. two-pronged institutional thoughtlessness; i.e. the way prisons roll on with little reference to the experiences, needs, and sensibilities of ageing individuals (Crawley 2005: 350). First, prison architecture is primarily designed for offenders who are young, energetic, and fit (e.g. cells and stairways are unfit for wheelchair patients or otherwise disabled prisoners), comprising the majority of prisoners (Kerbs and Jolley 2009; Potter et al. 2007). In addition, the prison regime and the services and programmes provided by the prison administration are poorly adapted to the needs of older prisoners (e.g. rigid timetable, sports and fitness infrastructure, outdoor exercise, prison labour, education, reintegration programmes, etc.) (see, e.g. Mann 2012; Crawley 2005; Crawley and Sparks 2005). Under the prevailing sameness principle, older prisoners have to adhere to the same rules and routines that apply to all prisoners (Crawley 2005: 355ff). By drawing on gerontological perspectives, it has been argued here that institutional thoughtlessness of prisons vis-à-vis older prisoners might as well be interpreted as a particular instance of the broader concept of ageism, whereby all prisoners are treated the same, despite differences in *age dimensions.*

## The Management of the Older Prisoner

In general, segregation is associated with the distance between distinct groups of people on the basis of certain characteristics. It has been argued that age segregation leads to ageism, fear, and a cycle of segregation-ageism (Hagestad 2008: 27). Ageism helps to deal with people's concerns about ageing—by *physically* and *psychologically* distancing older people (Fineman 2011, 57, own emphasis). Simone De Beauvoir wrote that 'they [old age adults] are not "us", they are "another being"' (De Beauvoir 1972: 3, own emphasis). In this sense, age segregation could be perceived as a specific expression of, and/or catalyst for,[14] *ageism*—a specific manifestation of 'othering' (cf. footnote for a brief explanation of this concept).[15] In applying this concept to older adults, Van Dyk (2016) noticed that older adults were increasingly excluded from society by radical 'othering' She discerned 'a double process of othering in later life' (2016: 9) with the young-old being valued as the 'other', and the oldest old being disdained as a different 'other'.

This distance vis-à-vis older adults is externalised through *spatial segregation* when individuals of different ages do not occupy the same space and hence cannot engage in face-to-face interaction (Hagestad 2008: 27). An extreme version of spatial age segregation occurs in intentionally age-homogenous housing and/or care arrangements, such as nursing homes, assisted living facilities, and retirement communities (Hagestad 2008: 27). Whilst this is a well-established tradition, it was not until fairly recently that questions arose with regard to penitentiary institutions spatially segregating its oldest prisoners.

In considering this, Michel Foucault's notion of 'heterotopia'[16] is useful. For Foucault, heterotopia denotes:

> *a signifier of spaces and places—geographical, architectural, institutional, and sometimes even purely symbolic—in which society defines, locates, and confines that which is 'its other'.* (Foucault, cited in Aronson 2013: 237)

Heterotopia is used as a philosophical lever here, to draw parallels between the underlying mechanisms behind the genesis of nursing homes and prisons. Foucault distinguishes between two major types: the heterotopia of 'crisis' and that of 'deviation' (Foucault 1967/1984). Heterotopias commonly materialise in closeted institutions familiar to modern political societies: the hospital, the nursing home, the prison, and the psychiatric ward. These are spaces in which 'individuals whose behaviour is deviant in relation to the required mean or norm are placed' (Foucault 1967/1984: 5). By drawing on this insight, parallels between nursing homes and prisons are deviant heterotopias. In this respect, both prisoners' and older adults' behaviour is deviant in relation to the respectively law-abiding and active, productive norm. However, Foucault has also argued that the nursing home is a transitional site between middle and old adulthood, or the transition between life and death, real and unreal, all of which occurs away from home and the public eye (cf. crisis heterotopia) (Foucault 1967/1984). Foucault identifies the retirement home as a heterotopia between deviance and crisis:

> ... since, after all, old age is a crisis, but is also a deviation since in our society where leisure is the norm, idleness is a sort of deviation. (Foucault 1967/1984: 5)

According to Foucault, heterotopias are almost invisible because they are conceived as 'natural' by society, but they are nevertheless the means of disciplining, controlling, and punishing the different and deviant. The extant literature identifies a wide range of strategies that have been utilised to manage older adult detainees in the prison environment (Baidawi et al. 2011: 5). It has been suggested that geriatric, or nursing home–type, programmes have increasingly determined the penal landscape to accommodate older prisoners with chronic health concerns and/or terminal illnesses, within a larger prison (Angus 2015).

## Segregation

There are instances of specialised prisons, designed to accommodate 'older prisoners', irrespective of chronic health concerns and/or terminal illnesses. Examples include the State Correctional Institution, Laurel Highlands, a geriatric and special needs facility in Central Pennsylvania and a 50-bed prison in Singen, Germany (Baidawi et al. 2011: 5). The latter example is considered a pioneer in Europe for dealing with ageing prisoners (Giele 2015: 12; von Borstel et al. 2014: 15). It is striking that the equipment inside this prison is not significantly different from ordinary prisons; the everyday life of these prisoners, however, looks very different (von Borstel et al. 2014: 15). Although the prison is designed as a closed facility, it is oriented as an *open prison regime*, with a daily routine that is less predetermined than in ordinary closed prison facilities (von Borstel et al. 2014: 15). Here, all prison cells are open between 7 am and 10 pm, and 24/7 in cases of multi-occupant cells without a toilet (Rennhak in von Borstel et al. 2014: 15). Tendencies like loneliness and social isolation in later life are counteracted by organising common shopping trips and hikes, whilst health support is offered in the form of age-appropriate sport programmes, discussion groups, and music and cooking classes (Schollbach and Krüger in von Borstel et al. 2014: 15). Extended visiting possibilities (six hours a month) and suitable occupations for those prisoners who are still required to work are notable differences from ordinary prisons and their regimes (Giele 2015: 13; von Borstel et al. 2014: 15). It should be noted, however, that not every 'older' German prisoner is eligible for admission to this institution. The prisoner candidate must meet the following eligibility requirements: (1) the prisoner must be male and with a chronological age of at least 62 years; (2) he must reside in Baden-Württemberg state; and (3) he must have been sentenced to serve at least 15 months.

Parallel to the institutions described above, other correctional facilities have established needs *units* or *wings* to cater for old(er) prisoner populations (cf. for units: Hayes et al. 2013: 5; Fry and Howe 2005).[17] The reported benefits of such units are similar to that of the 'nursing home-type' prisons (cf. supra), with centralised (health) care and resources reducing costs associated with staffing, medical care and transport, and age segregation of prisoners alleviating issues around prisoner victimisation, as well as enabling more targeted programming and rehabilitation efforts (Angus 2015; Kerbs and Jolley 2009).

An illustration of partial segregation is to be found in the 'Kornhaus', a special department for older inmates in the prison of Schwalmstadt in Hesse, Germany (von Borstel et al. 2014: 15). With a capacity for 61 prisoners who have reached the chronological age of 55 and classed as low risk, with regard to violence and absconding, older prisoners are transferred to this facility and mostly accommodated in single-occupancy cells (von Borstel et al. 2014: 15). Cells are unlocked at all times, and the facility offers medical, psychological, pedagogical, sports-pedagogical, and religious services (von Borstel et al. 2014: 15). Similar to the regime in Singen, residents of the Kornhaus have comparably generous visiting regulations (six hours a week) and can take part in memory training, age-appropriate sport groups, discussions, and information events on age-specific topics (Schollbach and Krüger in von Borstel et al. 2014: 15). The Kornhaus prison staff receive a special training programme that focusses on the special needs of older prisoners, and the prison functions on the basis of constant development (Roos and Eicke in von Borstel et al. 2014: 15). However, Von Borstel et al. (2014) noted that efforts to change conditions often result in only minor differences to the daily life of a prisoner, which are epitomised by the obsolete architecture and/or sentencing and correction policies and practices (von Borstel et al. 2014: 15f.). In the same vein, the prison administrations of Detmold, Bielefeld, Waldheim, and Bayreuth in Germany have installed special departments for older prisoners offering age-specific treatment and—in part—barrier-free facilities (von Borstel et al. 2014: 16).

With reference to the United States, Aday (2006: 216) identified a proliferation of projects for 'frail' and older inmates which offered unique programmes and services, including standalone facilities or secure nursing homes (cf. for an enumeration of the corresponding prison systems (Aday 2006: 216). In addition, there are a number of other states that group 50+ prisoners together in separate medical units, which might mix older prisoners with younger disabled ones (Aday 2006: 216). Aday points to the well-established practice of grouping prisoners of similar ages together in dormitory style cells with lower-level bunk beds (Aday 2006). Most specialised units are equipped with handrails, lower bunks on main-floor tiers, elevated toilets, and wheelchair accessibility, whilst prison-controlled thermostats, fluorescent lighting, strobe-lighted fire alarms, and non-slippery flooring surfaces are additional amenities (Aday 2006: 217).

## Integration

Whilst age integration in prisons commonly equates to the absence of special provisions for the oldest segment of the prisoner population, it does not necessarily exclude a specific approach. In addition to setting up special units or wings for older prisoners, there have also been a number of initiatives directed at older prisoners in (age-integrated) prisons, such as True Grit (US) (Harrison 2006), RECOOP, and Age UK (UK) (cf. also Mann 2012: 115).

*Tegel*, for example, the largest German prison for male prisoners in Berlin, adopts a different approach to ageing prisoners (von Borstel et al. 2014: 16). Rather than creating a segregated department for older prisoners, the prison accommodates its senior inmates with the general prison section but provides special age-appropriate treatment programmes (Kammerer and Spohr in von Borstel et al. 2014: 16). Occupational therapy, age-specific consultation hours, and computer classes have proven popular amongst old-age prisoners (von Borstel et al. 2014: 16).

The literature reveals many advocates for policies or national strategies for the provision of age specific regimes for older prisoners, yet:

> A requirement for every prison to have an older prisoner policy *detailing* age specific regimes would reduce the ability of prison governors to provide regimes which reflect the actual and specific needs of prisoners. (Ministry of Justice 2013: no. 6, own emphasis)

## Reviewing the Rationale

In most cases, older prisoners are separated from the general population for *medical* and *security* reasons rather than their 'age' (Aday 2003). However, age segregation has been advanced as the response to many of the healthcare delivery problems older prisoners experience (cf. Mann (2008: 126–134) as it would allow for the relocation of older prisoners from facilities that are largely oriented towards the care and needs of younger inmates for variant reasons.

The argument for separate facilities on grounds of cost-effectiveness is related to those made from a *security* standpoint (Aday 2003), on the basis that older inmates:

> generally pose less of a security threat than do younger violent offenders, and a large number can be safely housed in a minimum-security setting if medical care is available. (Morton 2001: 83)

Older prisoners, it is argued, show greater conformity to the prison rules and are thus easier to supervise because they are less likely to engage in troublesome or conflictual activities (Kerbs and Jolley 2009: 122; Goetting 1984b: 299; Rubenstein 1982: 292). For the overwhelming majority of older prisoners, costs could be saved by housing them in less expensive, age-segregated minimum-security correctional settings (Kerbs and Jolley 2009: 122). McShane and Williams (1990), however, suggest that there is much evidence to show older prisoners have disciplinary records that possess similar characteristics to their non-aged counterparts. They argue that such older prisoners may also require segregation facilities (McShane and Williams 1990: 210).

According to Bleeker (2003: 416), there are no reasonable arguments for providing special living arrangements for older prisoners, other than an expected increase in their health care needs. He argues that care and guidance are increasingly needed for younger adult detainees (Bleeker 2003: 416) and suggests that the strongest case is for improving prison health care in general. This raises the fundamental debate between equity and equality in the provision of healthcare services for (older) prisoners (cf. Fein 2005). It this sense, Williams et al. (2012) reported that:

> although experts generally agree that the phenomenon of a rapidly aging prisoner population needs to be [scrutinised, and respectively] addressed, the lack of standardised and comprehensive data, specifically about health care conditions and the costs associated with older prisoners, poses a problem in the implementation of evidence-based solutions to increase cost-effective, quality care. (Williams et al. 2012: 1477)

In response to the question, 'is there a case for a differentiated treatment of older male prisoners, for example, by accommodating them in separate older prisoner units', Chu (2016) conducted a small-scale qualitative inquiry into the experiences of older adult prisoners in England and Wales. He collected primary data from 16 formal interviews and 2 formal group discussions with older adult male prisoners, observations by 'hanging around', and informal conversations with staff (Chu 2016: 17ff.). Almost all his participants noted that they had never been physically bullied by a younger prisoner during any part of their custodial sentence (Chu 2016: 33). By contrast, most participants noted that they were often intimidated by their younger counterparts in the psychological sense (Chu 2016: 34). Almost all participants complained about the excessive noise in prison, invariably ascribed to the behaviour displayed by younger prisoners (Chu 2016: 34). Chu argues that this excessive loud noise and the regime's general inability to control it represented an additional and unnecessary pain of imprisonment, giving support for the segregation of older prisoners, at least to quieter sections of the prison (Chu 2016: 35).

The prison officers involved in Mann's (2008) study opposed the idea of age segregation:

> At the end of the day, the person who you've locked up for thirty years has been locked up because he killed somebody and that doesn't change just because they start pissing themselves… No one does knee jerk reactions like the prison service … how is it better? It's only better for that one individual… I can understand segregating for the sake of their safety but how far are we gonna break it down? Are we gonna have all Muslim units, all Jewish; in short, we'll have ghettos and isn't that what diversity tells us not to do?' (As cited in Mann 2008: 257)

Age-segregated housing can actually increase negative effects for older prisoners, as it encourages dependency, increases ageing, and promotes a feeling of imminent death amongst its inhabitants, which was evidenced by many of the prisoners in her sample who referred to such units as 'God's waiting room', and were fearful of ending up there (Mann 2008: 257). Moreover, the officers in her study point to the devastating effects of similar facilities, which is strongly reminiscent of the concept of learned helplessness,[18] a term coined by Seligman (Peterson et al. 1993):

> We had an old boy here a while ago, he came from Kingston, where he had been in an old persons unit and it had destroyed him. When he arrived here, he was very unsure about what to do because we said to him, 'You do it', whereas at Kingston, it's, 'You'll sit in that chair, you'll sing and clap when we say so', and it'd destroyed him, but over his time here, he actually developed back into a person who thought, and I think it can destroy people if they put them all together and treat them like oldies… In most older people, the body is frail but the brain isn't and that needs stimulating. (As cited in Mann 2008: 257)

## The Older Prisoner and the Prison Environment

Older prisoners housed in the mainstream prison population generally have much greater opportunities for prison employment than those in sheltered environments (Aday 2003). In a rare comparative study, Marquart et al. (2000) examined the sociodemographic characteristics, health conditions, and problems of daily living in two groups of older prisoners: integrated and segregated prisoners in the Texas prison system over the age of 50.[19] They found that older inmates in the special unit were more likely to report feelings of boredom and indifference, whereas older men in the mainstream population were more likely to work, be active, and engage in the institutional routine, leaving little time for boredom to take over (Marquart et al. 2000: 88ff.). The work connected the men from the latter group to the same routines as younger inmates; they walked, worked, dined, and visited the gym or the yard, the school library, or the chapel with younger men (Marquart et al. 2000: 89). The authors did not suggest that these men woke each day eager to go to work, or that hard work made them healthy (Marquart et al. 2000: 89), as evidenced by the response of one of the non-facility older prisoners who nonetheless stressed its significance:

> I've worked all my life and prison ain't no different. Now it's not like I get up and run to the job [prison construction]. I don't necessarily like to

> work. I've got to work to keep m'self busy, you know, occupied. At work I do my job and I also talk to different guys and this keeps me going. It makes the time go by fast. I get up, go to chow, go to work, go to chow, shower, go to chow, and go to my house [45 of 46 inmates lived in cells, on all three tiers, and one was assigned to a dorm]. It don't sound like much to you but the day they take me off the job they might as well move me into that old guy's dorm [the Geriatric Facility]. (60-year-old construction worker) (Cited in Marquart et al. 2000: 88f.)

Notwithstanding the assumption that this Geriatric Facility existed in an isolated environment, Marquart et al. (2000: 89) found support for specific facilities amongst both older groups: at least for those who cannot make it in the general prison population, for reasons of victimisation or declining health. Although for what were quite different reasons, both group members implicitly pointed to opposing sides of the same coin: the person-environment relationship; the fit between the individual and the situational characteristics (Sundberg 2004: 134). What became apparent, was a lack of fit between the mainstream prison environment and the needs of older prisoners (Howse 2003: 1; Moore 1989: 181; for further details on the environmental/functional mismatch, cf. Williams and Abraldes 2007: 65), over what was apparent amongst prisoners with 'gross' physical or mental infirmity (Allen 2003) (cf. supra on institutional thoughtlessness). Although this has been touched upon lightly (Howse 2003: 1), it has seldom been considered in terms of the physical or physiological dimensions of age.

The notion of 'boredom' in the context of the above is likely to occur when a person's capabilities are greater than the demands placed upon them, as opposed to the anxiety that arises when the environmental demands are too great (Sundberg 2004: 134). Recognising this, some have suggested that segregating older prisoners might contribute to their social isolation and boredom due to *the lack* of programmes (Williams and Abraldes 2007: 65).

In addition to its quantitative component, boredom is comprised of an increasingly prominent qualitative dimension which de Bree (2013) argued, 'it is not only 'doing anything' that matters, but also what you are doing and why' (dixit de Bree 2013: 17, own translation). Similarly, Baars

(2012) has been attentive to the risks of a dominant contemporary culture of rigid, or even wrongful, applications of the 'ageing actively' approach, which is epitomised by the 'meaningless and aimless activation' of older adults living in institutions (Baars 2012: 74). Continuing this theme, he concluded that whilst it is frowned upon within the micro-communities inhabited by older individuals, the act of disengaging from participation in society is desirable for some. Additionally, he critiques the model of care environments that has become increasingly guided by management models and standards (Baars 2012; 74, 29f.), within which professionals endeavour, or are even forced, to demonstrate their own indispensability (Gubrium and Wallace 1990: 139f.). As such, the meaning of daily life for older prisoners has received little scrutiny.

This discussion dovetails with the distinction between *situative* (imposed from the outside) and *existential* boredom (welling up from the inside) (de Bree 2013: 17; Hermsen 2013: 4; Svendsen 2005: 41f.). Although the boundaries between different forms of boredom are not always as sharply drawn since they can overlap or feed into one another (de Bree 2013: 17), it is argued that situative boredom occurs when there is 'a longing for something that is desired', whereas existential boredom comprises 'a longing for any desire at all' (Svendsen 2005: 42). The former is aroused by something specific, like waiting for someone, or taking the train, whereas for the latter, the soul is without content and the world is neutral (Svendsen 2005: 41f.). It is the temporary tedium that results from a lack of engaging or pleasurable stimulation from the environment; put simply it is 'having nothing to do' that generates boredom (Barrett and Martin 2014: Chap. 12). By contrast, existential boredom is a more chronic feeling of emptiness, a paralysis of agency in which one's reality becomes meaningless (Eastwood et al. 2012).

According to Barbalet (1999), this emotion 'registers an absence of meaning' and could account for the prevalence of intergroup conflict and risk-taking behaviour (e.g. bullying, violence, drug taking, and racketeering) as potential releases from the boredom prisoners encounter (Barbalet 1999: 631). It could be argued that whilst this might have an impact on situational boredom, it might be considered less significant for their existential boredom, which transcends the specific environment or situation. However, there is some evidence to suggest that situative boredom has the potential to develop into something more like existential boredom (Barbalet 1999).[20]

Similarly, a respondent interviewed by Weiss and Bass (2002) voiced his concern about 'a life of drift' during old age, one of trying for a pleasant day and evening and not much more (Weiss and Bass 2002: 6). They describe drift as 'letting things happen, deciding on activities one at a time and so establishing a routine that is good enough, but not at all chosen' (Weiss and Bass 2002: 5). In this respect, Baars suggests that these individuals might simply offer little counterweight to the prevalent discourses and structures that are evident when someone feels ill or vulnerable (Baars 2012: 74).

## Intergenerational Conflict

Studies have raised questions as to whether (chronological) age is negatively related to victimisation across the entire life course and have provided some insights into the relationship between chronological age and inmate-on-inmate victimisation. Kerbs and Jolley (2007) suggest that older prisoners' social status, offence histories, and medical histories may play a role in different levels of victimisation (Kerbs and Jolley 2007: 191f.).[21] A higher social status in the general prison population may have been afforded once, because of their typically vast experience with crime and/or prison (Goetting 1985: 18), and that may well be disappearing (Kerbs and Jolley 2007: 191). Others have provided a more differentiated picture of the inmate hierarchy as one based *partially* on seniority, whereby recently admitted prisoners occupy the lowest place, and 'old cons' with long prison experience regulate norms and occupy leadership roles (cf. McCleery in Goetting 1985: 18). Although the effects of age and prison experience have not been sufficiently disentangled (Kerbs and Jolley 2007: 190f.), it is arguable that different subgroups of older prisoners may have different experiences that affect their safety and victimisation (cf. the literature on the typology of older prisoners based on sentencing histories; e.g. Goetting 1984a). As to the nature of their offences, there is some evidence that sex offenders, and paedophiles in particular, are more vulnerable to inmate-on-inmate victimisation (Kerbs and Jolley 2007: 191f.). Finally, their (younger) counterparts may see older prisoners' advanced age and age-related disability and morbidity as decreasing potential resistance to victimisation, especially for property crimes (Kerbs and Jolley 2007: 192).

Kerbs and Jolley (2007: 198–209) have added nuance to the extant scholarship by identifying the following four *subtypes* of victimisation amongst 65 older male prisoners (50+), listed in descending order of prevalence: psychological, property, physical, and sexual victimisation. They found that most (but not all) older prisoners remained silent or responded in a nonviolent matter to victimisation (cf. for reactions of prisoners to victimisation: Kerbs and Jolley 2007: 199; 202: such as seeking assistance from others (staff or peers) or walk away). As to the most common subtype of victimisation, i.e. psychological, wherein line cutting was the most common form, an additional distinction is drawn between 'insults' and 'labels'. The former was denoted as:

> pejorative and/or denigrating comments, which were typically proffered by younger prisoners with ageist attitudes', whereas the latter entails 'pejorative terms that are used publicly to identify a given older prisoner as fitting into one of three unique and publicly despised categories: snitches, sex offenders, and/or homosexuals ... with greater potential for physical victimisation'. (Kerbs and Jolley 2007: 200)

Although they have included two environmental concerns (exposure to excessive noise and second-hand cigarette smoke) under the term 'psychological victimisation' (Kerbs and Jolley 2007: 201f.), it is more generally discussed under the notion 'institutional thoughtlessness' (cf. supra).

An analysis of 262 older prisoners' (50+, North West England) responses to open-ended questions concerning the advantages and disadvantages of being an older person in prison showed many disadvantages pertained to mixing with the younger group, the noisy environment, and bullying (Hayes et al. 2013: 5).[22] Victimisation in the form of property theft was a common event, in part because older prisoners appear to be generous and therefore vulnerable to con games or are unable to protect their belongings due to sensory impairments (Kerbs and Jolley 2007: 202–205). In total, 29% reported being cheated or conned out of money by another inmate in the past year, and about 28% had property stolen by another inmate (Kerbs and Jolley 2007: 206).

The subtype of physical victimisation was found to be less common in comparison to psychological and property victimisation, and this included bullying behaviour (pushing, often at the hands of younger prisoners) more often than serious assaults, prompted by the opportunity to steal, or the escalation of interpersonal conflicts (Kerbs and Jolley 2007: 206f.). Nonetheless, it must be noted that some respondents were seriously injured and came close to losing their lives (Kerbs and Jolley 2007: 207). Although sexual inmate-on-inmate victimisation appeared to be negatively related to age, older sex offenders seemed more likely to report both minor and serious events of sexual victimisation (Kerbs and Jolley 2007: 208f.).

The common denominator in these types of victimisation was the assertion of (*mostly* by younger prisoners') dominance over older prisoners and the desire to control and dominate public spaces and resources within prison (Kerbs and Jolley 2007: 198; 200; 206). Younger prisoners tend to be more reliant on aggressive precautionary behaviours (Kerbs and Jolley 2009: 129), whilst older prisoners tend to employ 'passive precautionary behaviours': keeping to oneself, avoiding certain areas of the prison, spending more time in one's cell, avoiding activities (HRW 2012: 59–60; Kerbs and Jolley 2009: 129), and avoiding the inmate culture, where they may be easy and obvious targets (Kerbs and Jolley 2009: 130). Within a framework of desistance, Kerbs and Jolley (2009: 130) consider passive precautionary behaviour and avoiding strategies in combination with the disengagement from the hostile and violent inmate culture as an additional argument supporting age-segregated living arrangements. According to Kerbs and Jolley (2009), older prisoners may be engaged in a meaningful process of desistance, which might be exacerbated by being immersed in a social environment where ongoing criminal activity is held in high regard (Kerbs and Jolley 2009: 130). In sum, these authors draw on similar arguments which have been applied in establishing age-segregated living arrangements for juveniles: i.e. fear for their physical safety, identity development and social trajectories (Kerbs and Jolley 2009: 130).

## Intergenerational Ties

The contra opinion advocates the integration of older adult inmates with younger peers. Some, mostly prison officials, have claimed that the presence of older individuals amongst the general prison population elicits a calming or stabilising effect on the prison population (House of Commons Justice Committee 2013, HC 89-I: 22; Jurkowski 2008: 352; Rubenstein 1982: 292). Others have questioned this, by drawing upon examples (in age-integrated prisons) of jailbreaks, prison riots, the increasing numbers of first-time offenders in the older age group, and the fact that officials refer to inmates between the ages of 35 and 50 being involved (Ikuteyijo and Rotimi 2009; Morton 1992). Similarly, it has been argued that older prisoners exhibit:

> fewer problems with correctional staff, demonstrate more respect, commit fewer role infractions and in general get along better with other inmates. (Cited in Aday 2003: 147)

For Hayes et al. (2013: 3f.), a significant proportion of his sample indicated the *advantages* of being an older adult in prison: attitudes of staff, availability of old-age services, being respected by other prisoners and staff, and being able to help other prisoners (cf. Giele 2015; Gubler 2006; Zoutendijk 2007, 2009 on the role of (some) older prisoners towards youngsters). In a small-scaled Dutch study, Zoutendijk (2009) concluded that the majority of older inmates (65+) actually appreciated being amongst youngsters, despite the noise and the hectic pace (Zoutendijk 2009: 262). Likewise, Chu (2016: 35) concluded that the minority of participants who objected to being housed in an older prisoner unit tended to be younger and healthier and were able to interconnect with younger people.

The next argument relates to the relationship between generativity and the process of desistance hinted at above by Maruna (2001). Older prisoners, it is argued, can facilitate or promote desistance amongst younger prisoners by inter alia: warning them off a lifestyle of crime (Cash 2004: 149). This is a clear instance of how intergenerational ties are suggested to benefit both older and younger individuals in prison. However, in this line of reasoning, they are also being used as a means to another person's ends. Recently, segregation has been linked to the notion of generativity (Hagestad and Herlofson 2007: 351). Equally, a working paper of Moran, where he examines the notion of 'generativity', denoted here as 'a concern for and commitment to promoting development and wellbeing of future generations' (cf. McAdams 2013: 31), within a rehabilitation programme entitled 'Council for Unity' at a county jail in Long Island. By drawing on data from participant observation and qualitative interviews, Moran (2014) found evidence of a specific generative role for older[23] prisoners. A generative desire that existed prior to the programme and constituted the most frequently expressed reason for participation amongst older inmates (Moran 2014: 13). The group had been deliberately mixed so as to provide opportunities for positive mentorship and Moran's hypothesis was tentatively borne out by his findings where older participants engaged in generative actions, often through what Moran has referred to as 'mini-speeches'.[24] In the context of desistance, Moran points to the 'mixed' nature of their role as mentors in the prison setting; not only do they constitute 'positive role models by virtue of their commitment to changing and their endorsement of the value of desistance', but they also provide 'living examples of the "feared self" in terms of an undesirable future existence' (Moran 2014: 14). Perhaps more importantly, their generative acts towards younger participants may assist older offenders in their own reform efforts, in a three-tiered mode: (1) it provides them with a sense of purpose (thereby reforming the condemnation script to the generative script)[25]; (2) it enables them to come to terms with past offending (i.e. redemptive or atoning quality); and lastly, (3) it allows them to experiment with a socially legitimate role of which they could claim no small competency (thus could provide the same sense of empowerment and potency they were seeking in a criminal lifestyle) (Moran 2014: 14–16; Maruna 2001).

## Bifurcation

In an endeavour to resolve this debate, Swiss researchers conducted interviews amongst 35 older prisoners (prison population segment aged 50+) from Switzerland and 40 stakeholders from three European countries divided into three groups: healthcare professionals or researchers; prison administrators or policymakers; and scholars working with non-governmental or international organisations. The purpose was to capture the perspectives of people living and working in a prison setting on the living conditions of the oldest segment of the prisoner population (Wangmo et al. 2017).[26] The findings of their study confirmed the highly bifurcated nature of the extant scholarship and the debates: almost half of their respondents in each group (both prisoners and professionals) made the case for (age)segregation, whereas the other half advocated 'mainstreaming' the older prisoner population (Wangmo et al. 2017: 681). Crawley drew, the similar conclusion that 'neither prisoners nor staff are unequivocally supportive of the creation of distinct age-segregated units' (Crawley 2005: 360). The advocates for consolidated housing put forward the argument that 'prisons should mirror society' and regarded it as paramount to retain opportunities for generational exchange and that 'ghettoisation', or spatial marginalisation, must be discouraged (Wangmo et al. 2017: 681–683).

Some older prisoners explained their role as advisors within the prison setting and suggested that they enjoyed sharing their experiences with younger prisoners, whilst like-minded stakeholders emphasised the responsibility and help from younger prisoners to their older counterparts (Wangmo et al. 2017: 682). Most of those who supported the segregation of older prisoners recommended a separate *unit* (or wing) within the prison, rather than a separate institution for this group (Wangmo et al. 2017: 683). It was argued that a separate unit would allow for continuity whilst providing well-needed adjustments for ageing prisoners' health care and their environmental needs (Wangmo et al. 2017: 683ff.). The participants highlighted, amongst other things, better medical supervision (i.e. the presence of a doctor and regular check-ups) but also self-care (Wangmo et al. 2017: 684f.). Moreover, most segregation-oriented participants preferred grouped living arrangements where tranquillity, space, more freedom to make small choices, and better adapted reinforced rooms (Wangmo et al. 2017: 684f.).

Aside from spending time with younger peers in the prison workplace (Wangmo et al. 2017: 685), it was not specified, however, what form this interplay with the mainstream population should take. Most participants specified that a chronological eligibility cut-off age of 50 was too low and that admission to segregated housing should be based on consent (Wangmo et al. 2017: 686f.). Part of the sample supporting the separation of prisoners by age group suggested that it would be preferable to segregate prisoners not only on age per se but also on the type of sentence; security level and likelihood of release; and/or to cluster prisoners on their capabilities (Wangmo et al. 2017: 685–689). It is striking that violence against older prisoners, the pivotal argument in support of age-segregated living situations in the predominant literature, was not borne out by this study (Wangmo et al. 2017: 688). The authors attribute this to the scarce opportunities for violence in the examined context of a small-scale prison construction in Switzerland, and that the sampled prisoners were housed in a single cell. It is also possible to consider potential bias due to the sensitivity of 'victimisation' or its relationship to other events.

Zoutendijk (2007: 61ff, 2009: 249) reached a similar conclusion in his exploratory qualitative study[27] in the Netherlands. In this case, most of the participants did indicate that they felt safe within the penitentiary institution. However, there evidence suggested a link with sexual offences increased the likelihood of being both psychologically and physically victimised (Zoutendijk 2007: 61f.), which gives some support to Kerbs and Jolley's (2007: 191f.) assertion that sex offenders are more vulnerable to inmate-on-inmate victimisation. Ageing sex offenders have consequently been dubbed as 'dually disadvantaged' with stigmatised offence and advanced age (Kerbs and Jolley 2007: 213). Moreover, old age in itself might give rise to the suspicion that one is a sex offender. Some participants who had been imprisoned for a non-sexual offence faced the suggestion that they either had been or were serving time for a similar type of offence revealed similar levels of vulnerability (Zoutendijk 2007: 61f.; 63).

The current emphasis on *age segregation* may therefore give rise to confusion. Although sex offenders are overrepresented in the older prisoner population, it would be incorrect to assert that both groups coincide. Not only does this spurious relationship lead to stigmatisation, it could also undermine and impact policy support for issues of the entire older prisoner population. Snacken's (2001) contribution is of particular interest here. She describes how two systems coexist in Belgian prisons: segregation and concentration of sexual offenders in 'vulnerable prisoners' units' and controlled integration (Snacken 2001: 472). Snacken's findings suggested that segregation reinforces stigmatisation and that every contact prisoners have outside the segregated unit posed a risk of violence. Whereas in integrated prisons, violence towards sex offenders appeared to be fairly limited, provided prison staff and prison governors were consistent in imposing disciplinary measures for such assaults (Snacken 2001: 472).

Although it is argued that the safety of older prisoners would be significantly enhanced by age-segregated facilities, Kerbs and Jolley (2009: 132) urge prison regimes to be wary of substituting one form of victimisation (inmate-on-inmate victimisation) with another (staff-on-inmate victimisation and/or programmatic neglect). There is much evidence to suggest that age-segregated facilities have perpetrated this kind of institutional maltreatment.

Arguing against segregated accommodation for prisoners, the establishment of age-segregated facilities might also give rise to 'iatrogenic' problems. These include the placement of older prisoners in remote, poorly resourced facilities, with a lack of adequate access to programming and (sufficient) competent medical staff, particularly problematic for inmates with complex bio-psycho-social problems (Kerbs and Jolley 2009: 132). This has prompted Morton (2001), for example, to advocate for mainstreaming older inmates in an effort to 'help insure that older inmates are not denied access to programs and services available to other inmates' (Morton 2001: 83).

Goetting (1985) summed up the potential adverse effects of both segregation and having special policies for older prisoners by highlighting some of the corresponding dilemmas:

> Segregation can reduce pressure for standardized treatment, allowing stereotyping and, in the case of the elderly, small numbers to influence institutional policy. And as has been shown to be the case with women, such influence is not always to the advantage of the minority group. In other words, segregation is conducive to discrimination. The question of appropriateness of segregation of adult prisons by race and age, and of providing special policies and treatment for the elderly poses a true dilemma to those professionals who are concerned with the well-being of prisoners as well as to those interested in prison policy and administration. While human needs must be accommodated, it is important that care be taken to avoid the discrimination that can contaminate a dual prison system. While it may be a fine line that separates special need satisfaction and discrimination, it is incumbent upon corrections administrators to discover that line, and to create and implement policy accordingly. Perhaps our best conceptual approach to the consideration of minorities in prison is to their consideration in society at large. If problems of minorities in this country were ever to be alleviated, problems of their representatives in prison and in other institutions, for that matter, would follow suit. (Goetting 1985: 20ff)

In considering the idea of segregating distinct categories of prisoners from the mainstream population, Foucault's notion of *heterotopia* is especially useful in negotiating parallels between disciplinary institutions and latent knowledge about norms in society. A 'Foucauldian approach' to this inquiry offers a potentially innovative methodological route to revealing more nuanced understandings of the meanings attached to segregating categories of prisoners ('segregation within segregation') and unpicking the underlying assumptions and effects.

Whilst an emerging body of research places a growing emphasis on the experiences of people who grow old(er) in the prison environment, it lacks the characteristics of a coherent field of research. Very few empirical studies have produced a distinct understanding of the debate surrounding segregation or integration of the older prisoner (Chu 2016: 14). It remains unclear whether or not there are significant differences in the experiences of older adults who are separated and those subject to consolidated housing (Chu 2016: 14). These underlying ambiguities have given impetus to this inquiry into the experience of imprisonment in later life within an integrated and segregated prison environment.

# Notes

1. For a more extensive discussion of the chronological time concept, see Dannefer and Phillipson (2010: 368–369) and Baars (2007: 4–7).
2. For a detailed reference to its 'undeserved prominent role', see, e.g. Baars (2007: 2–6).
3. For a critical discussion on the use of biomarkers of aging to formulate biological age, see Arking (2006: 91).
4. In light of recent developments, it can be noted that more technologically advanced communication systems might affect prisoners' functional age to communicate in terms of technological (il)literacy (see, for instance, the PrisonCloud IT platform (e-BO Enterprises 2018), which has been already introduced in the Belgian prison system). The few data we have gathered in relation to this subject might lead us to suggest that such technological advancements might have an impact on older adults in two directions: such an approach might be beneficial for some older prisoners yet disadvantageous for others.
5. 'Porter' is an institutional job title for prisoners who are assigned to any prison function (Dick et al. 2016: 104).
6. She has drawn on in-depth interviews with male prisoners over 50 years of age who have spent at least 10 years in prison and are housed a large prison in the southeastern United States (N = 19, M = 60 yrs).
7. We still use the term 'retirement' to differentiate between those prisoners who engage in paid work in the prison environment and those who do not.
8. These authors have drawn on 22 narrative interviews with 'older' male prisoners (i.e. denoted as having a chronological age of ≥60) in three Swiss penitentiary institutions.
9. See Biggs (2005) for an outline of the different theories on old age identity.
10. The term prison regime should be understood in the same way as in the study of Crawley and Sparks, also subsequently employed by Mann, to incorporate the timetables, physical layout, practices, rules, and activities of the prison establishments (Mann 2012: 41; Crawley and Sparks 2005: 352).
11. Butler describes how his personal background made him more susceptible to the prejudice against older people: '*I was aware of ageism before it had a name, conscious as I was of being different from other boys by virtue of*

*having older "parents" who, in fact, were my grandparents*' (Butler 2005: 84, own emphasis). His first encounter with age prejudice was as a medical student where he was affected by the cruel medical lexicon that reflected negative attitudes towards patients, especially older patients (Butler 2005: 84). During the 1950s–1960s, he worked in nursing homes and was deeply involved in research, during which he became aware of the inaccuracy of ideas about the process of ageing and older people, and noticed how older persons in facilities, as well as their poverty, were made invisible (Butler 2005: 84f.). It was his personal observation of strong protest of many neighbours against city plans to set up housing for the older people which led him to coin the term 'ageism' (Butler 2005: 85).

12. Read for the gains of positive and negative ageism: Palmore (1999: 120).
13. Although institutional ageism occurs within and between institutions in society, the focus lies here on ageism within prison institutions.
14. Age segregation has been identified both as an antecedent and a consequence of ageism (cf. Dykstra and Fleischmann 2016: 2; Vanderbeck 2007: 209).
15. For an examination of De Beauvoir's description of aged as the other, cf. Stoller (2014: 18). Othering is a concept that denotes the positioning of minority and less powerful groups as inferior to dominant groups, and otherness is thus a marker of difference (Zevallos 2011).
16. The term 'heterotopia' is derived from the Greek words for '(an)other' (*heteros*) and 'place' (*topos*) and thus literally means 'other place' (Filimon 2009: 17).
17. Examples of units housing aged inmates can be found in, inter alia, England (e.g. HMP Frankland, HMP Wymott, HMP Norwich) (Mann 2012: 115) and Australia (Kevin Waller Unit at Long Bay Correctional Complex) (Baidawi et al. 2011).
18. This refers to the maladaptive passivity amongst humans as a result of their experience of a lack of control over their environment and events.
19. They conducted structured interviews amongst 23 inmates from the Geriatric Facility at the Estelle Unit and amongst 46 inmates who were housed elsewhere in the Estelle Unit. Estelle constitutes a maximum-security unit in the Texas prison system (Marquart et al. 2000: 81). The Geriatric Facility is a single-level dormitory-style detached and separate unit within the greater prison compound to which inmates are assigned to upon request or because of medical necessity

(Marquart et al. 2000: 81). The data collection also includes brief interviews with seven nurses and one physician who worked closely with inmates housed in this facility (Marquart et al. 2000: 82).

20. See the phenomenological study of Hanneke van der Meide into the experiences of older hospital patients, where what initially starts as 'situational/common boredom' slowly changes into 'existential boredom' for Mrs. Bloem (85 yrs, admitted to hospital) (van der Meide and Olthuis 2013: 31). Svendsen comments that different people are afflicted by boredom to differing degrees. He continues his argument by stating that although everyone is affected by boredom sooner or later, it is when 'boredom strikes hard' that one questions his entire existence (Svendsen 2005: 153).
21. Cf. in this regard also the interview with chief medical officer at California Medical Facility in California wherein he states that urinary and faecal incontinence could provoke ostracisation and even physical assault from annoyed younger inmates in dormitory environments (HRW 2012: 59, footnote 123).
22. 'Difficulty getting around prison' and 'loss of freedom' were two remaining disadvantages (Hayes et al. 2013: 4).
23. The author of this paper does however not provide the reader with a definition of 'older'. Although his reference to Erikson's positioning of 'generativity' in middle/late adulthood would lead us to suggest that this author *also* includes prisoners who are chronologically much younger than is demarcated in this inquiry, his findings remain valid and fundamental when applied to the mechanisms of ageing.
24. See for an example of such generative contribution to a group discussion, Moran (2014: 13f.).
25. See for the origin of these two types of personal narratives, i.e. a 'condemnation script', favoured by active offenders, and a 'generative script', favoured by desisters (Maruna 2001).
26. Their study is part of the larger scale project 'Agequake in Prisons', which is aimed at understanding the health and ageing experiences of older prisoners in Switzerland (Wangmo et al. 2017: 678). More detailed information is to be found on the website of the Swiss National Science Foundation: http://p3.snf.ch/Project-135035 (Accessed on June 5, 2017).
27. In addition to penitentiary files and conversations with chaplains, he conducted interviews amongst 21 older inmates (of which one female),

dispersed over various penitentiary institutions in the Netherlands (Zoutendijk 2007: 12ff.).

# References

Achenbaum, A. W. (2015). A History of Ageism Since 1969. *Generations, 39*(3), 10–16.

Aday, R. H. (1994). Golden Years Behind Bars: Special Programs and Facilities for Elderly Inmates. *Federal Probation, 58*(2), 47–54.

Aday, R. H. (2003). *Aging Prisoners: Crisis in American Corrections*. Greenwood Publishing Group.

Aday, R. H. (2006). Managing the Special Needs of Ageing Prisoners in the United States. In A. Wahidin & M. Cain (Eds.), *Ageing, Crime and Society* (pp. 210–229). Willan Publishing.

Allen, D. (2003). Prisoners' Health: Key Concerns. *Nursing Older People, 15*(7), 6.

Allersma, J. (1970). Bejaardencriminaliteit. *Nederlands Tijdschrift voor Gerontologie, 12*, 69–85.

Andersson, L. (2002). *Cultural Gerontology*. Greenwood Publishing Group.

Angus, C. (2015). *Older prisoners—Trends and challenges.pdf* (E-Brief EB 14/2015; p. 16). NSW Parliamentary Research Service. Retrieved from https://www.parliament.nsw.gov.au/researchpapers/Documents/older-prisoners-trends-and-challenges/Older%20prisoners%20-%20trends%20and%20challenges.pdf.

Antonucci, T., & Akiyama, H. (1997). Multiple Paths of Midlife Development. In M. E. Lachman & J. B. James (Eds.), *Multiple Paths of Midlife Development* (pp. 145–170). University of Chicago Press.

Arking, R. (2006). *Biology of Aging: Observations and Principles: Observations and Principles*. New York: Oxford University Press.

Aronson, O. (2013). In/visible Courts: Military Tribunals as Other Spaces. In D. Cole, F. Fabbrini, & A. Vedaschi (Eds.), *Secrecy, National Security and the Vindication of Constitutional Law* (pp. 229–248). Edward Elgar Publishing.

Baars, J. (2007). Chronological Time and Chronological Age: Problems of Temporal Diversity. In J. Baars & H. Visser (Eds.), *Aging and Time* (pp. 1–13). New York: Baywood Publishing Co.

Baars, J. (2009). Problematic Foundations: Theorizing Time, Age, and Aging. In V. L. Bengtson, D. Gans, N. M. Putney, & M. Silverstein (Eds.), *Handbook of Theories of Aging* (2nd ed., pp. 87–100). Springer Publishing Company.

Baars, J. (2012). Het nieuwe ouder worden. Paradoxen en perspectieven van leven in de tijd (3rd ed.). SWP Uitgeverij.

Baidawi, S., Turner, S., Trotter, C., Browning, C., Collier, P., O'Connor, D., & Sheehan, R. (2011). Older Prisoners—A Challenge for Australian Corrections. *Trends & Issues in Crime and Criminal Justice, 426*, 1.

Barak, B., & Schiffman, L. G. (1981). Cognitive Age: A Nonchronological Age Variable. *ACR North American Advances, NA-08*. Retrieved from http://acrwebsite.org/volumes/5867/volumes/v08/NA-08.

Barbalet, J. M. (1999). Boredom and Social Meaning. *The British Journal of Sociology, 50*(4), 631–646. https://doi.org/10.1111/j.1468-4446.1999.00631.x.

Barrett, E., & Martin, P. (2014). *Extreme: Why Some People Thrive at the Limits*. Oxford University Press.

Baumeister, B., & Keller, S. (2011). *Alt werden im Straf- und Massnahmenvollzug*. ZHAW Soziale Arbeit.

Becker, E. (1973). *The Denial of Death*. Free Press.

Biggs, S. (1997). Choosing Not to Be Old? Masks, Bodies and Identity Management in Later Life. *Ageing & Society, 17*, 553–570.

Biggs, S. (2005). Beyond Appearances: Perspectives on Identity in Later Life and Some Implications for Method. *The Journals of Gerontology Series B: Psychological Sciences and Social Sciences, 60*(3), S118–S128. https://doi.org/10.1093/geronb/60.3.S118.

Bleeker, J. A. C. (2003). Ouderen met psychiatrische stoornissen in detentie. In T. I. Oei & M. S. Groenhuijsen (Eds.), *Actuele ontwikkelingen in de forensische psychiatrie* (pp. 411–422). Kluwer.

von Borstel, M., Burkhardt, S.-U., & Graebsch, C. (2014). *Vulnerable Groups of Inmates. Country report—Germany* (Re-socialisation of Offenders in the EU: Enhancing the Role of the Civil Society (RE-SOC)) (p. 36). University of Applied Sciences and Arts. Retrieved from https://www.dadinternational.org/images/VulnerableGermany.pdf

de Bree, M. (2013). Over handen wassen, verveling en patiëntveiligheid. *Podium voor Bio-ethiek, 20*(2), 16–18.

Butler, R. N. (1989). Dispelling Ageism: The Cross-Cutting Intervention. *The ANNALS of the American Academy of Political and Social Science, 503*(1), 138–147. https://doi.org/10.1177/0002716289503001011.

Butler, R. (2005). Ageism: Looking Back over My Shoulder. *Generations, 29*(3), 84–86.

Butler, R. N. (2008). Ageism. In E. A. Capezuti, E. L. Siegler, & M. D. Mezey (Eds.), *The Encyclopedia of Elder Care: The Comprehensive Resource on Geriatric and Social Care* (2nd ed., pp. 25–26). Springer Publishing Company.

Carmichael, C. L., Tsai, F.-F., Smith, S. M., & Caprariello, P. A. (2007). The Self and Intimate Relationships. In C. Sedikides & S. J. Spencer (Eds.), *The Self* (pp. 285–310). Psychology Press.

Cash, A. (2004). *Psychologie voor Dummies*. Pearson Education Benelux.

Chasteen, A. L., & Cary, L. A. (2015). Age Stereotypes and Age Stigma: Connections to Research on Subjective Aging. In M. Diehl & H.-W. Wahl (Eds.), *Annual Review of Gerontology and Geriatrics*. Springer Publishing Company.

Choi, N. G., DiNitto, D. M., & Kim, J. (2014). Discrepancy Between Chronological Age and Felt Age: Age Group Difference in Objective and Subjective Health as Correlates. *Journal of Aging and Health, 26*(3), 458–473. https://doi.org/10.1177/0898264314523449.

Chu, V. (2016). *Greying Behind Bars: The Older Male Offender's Experience of Prison Life and Preparations for Resettlement*. Master dissertation, University of Cambridge. Retrieved from http://howardleague.org/wp-content/uploads/2017/02/Greying-Behind-Bars.pdf.

Clemmer, D. (1966). *The Prison Community*. Holt.

Crawley, E. (2005). Institutional Thoughtlessness in Prisons and its Impacts on the Day-to-Day Lives of Elderly Men. *Journal of Contemporary Justice, 21*(4), 350–363.

Crawley, E., & Sparks, R. (2005). Hidden Injuries? Researching the Experiences of Older Men in English Prisons. *The Howard Journal, 44*(4), 345–356.

Dannefer, D., & Phillipson, C. (2010). *The SAGE Handbook of Social Gerontology*. London: SAGE Publications.

De Beauvoir, S. (1972). *The Coming of Age*. G.P Putnam.

Deaton, D., Aday, R. H., & Wahidin, A. (2009). The effect of health and penal harm on aging female prisoners' views of dying in prison. *Omega, 60*(1), 51–70.

Dick, A. J., Rich, W., & Waters, T. (2016). *Prison Vocational Education and Policy in the United States: A Critical Perspective on Evidence-Based Reform*. New York: Springer Nature.

van Dyk, S. (2016). The Othering of Old Age: Insights from Postcolonial Studies. *Journal of Aging Studies, 39*, 109–120. https://doi.org/10.1016/j.jaging.2016.06.005.

Dykstra, P., & Fleischmann, M. (2016). Are Societies with a High Value on the Active Ageing Index more Age-Integrated? In A. Zaidi, S. Harper, K. Howse, G. Lamura, & J. Perek-Bialas (Eds.), *Building Evidence for Active Ageing Policies: Active Ageing Index and its Potential*. Palgrave Macmillan. Retrieved from https://repub.eur.nl/pub/79458/.

Eastwood, J. D., Frischen, A., Fenske, M. J., & Smilek, D. (2012). The Unengaged Mind: Defining Boredom in Terms of Attention. *Association for Psychological Science, 7*(5), 482–495.

e-Bo Enterprises. (2018). *PrisonCloud*. Retrieved from https://ebo-enterprises.com/prisoncloud.

Farber, M. L. (1944). Suffering and time perspective of the prisoner. In K. Lewin, C. Meyers, J. Kalhorn, M. Farber, & J. French (Eds.), *Authority and Frustration: Studies in Topological and Vector Psychology, III* (pp. 153–227). University of Iowa Press.

Fazel, S., Hope, T., O'Donnell, I., Piper, M., & Jacoby, R. (2001). Health of Elderly Male Prisoners: Worse than the General Population, Worse than Younger Prisoners. *Age and Ageing, 30*(5), 403–407.

Fazel, S., Hope, T., O'Donnell, I., & Jacoby, R. (2004). Unmet Treatment Needs of Older Prisoners: A Primary Care Survey. *Age and Ageing, 33*(4), 396–398. https://doi.org/10.1093/ageing/afh113.

Featherstone, M., & Hepworth, M. (1991). The Mask of Ageing and the Postmodern Lifecourse. In M. Featherstone, M. Hepworth, & B. S. Turner (Eds.), *The Body: Social Process and Cultural Theory* (pp. 371–389). SAGE.

Fein, R. (2005). On Achieving Access and Equity in Health Care. *The Milbank Quarterly, 83*(4). https://doi.org/10.1111/j.1468-0009.2005.00427.x.

Festinger, L. (1954). A Theory of Social Comparison Processes. *Human Relations, 7*(2), 117–140. https://doi.org/10.1177/001872675400700202.

Filimon, E. C. (2009). *Worlds in Collision—Angela Carter's Heterotopia*. GRIN Verlag.

Fineman, S. (2011). *Organizing Age*. Oxford University Press.

Fiselier, J. P. S. (2007). Oud (worden) in de gevangenis. In B. F. Keulen, G. Knigge, & H. D. Wolswijk (Eds.), *Pet af: Liber amicorum D.H. de Jong* (pp. 43–62). Wolf Legal Publishers. Retrieved from https://www.rug.nl/research/portal/files/14458168/05_oud.pdf.

Flynn, E. E. (1992). The Graying of America's Prison Population. *Prison Journal, 72*, 77.

Foucault, M. (1975). *Discipline, toezicht en straf. De geboorte van de gevangenis* (Vertalerscollectief Historische Uitgeverij, Vert). Éditions Gallimard/Historische Uitgeverij.

Foucault, M. (1984). Des Espace Autres. Heterotopies. (J. Miskowiec, Vert.). *Architecture, Mouvement, Continuité, 5*, 46–49.

Foucault, M. (1989). *Surveiller et Punir, Naissance de la prison*. Gallimard.

Fry, D., & Howe, D. (2005). Managing Older Prisoners at HMP Wymott. *Prison Service Journal, 160*, 11–13.

Giele, D. (2015). *Vergrijzing van de gevangenispopulatie. Is een gevangenis voor oudere gedetineerden de toekomst voor België?* Master dissertation, Universiteit Gent. Retrieved from http://www.scriptiebank.be/sites/default/files/webform/scriptie/Giele_Doris_masterproef2015_1.pdf.

Glover, B. M. (2012). Accommodating Older and Disabled Prisoners in England and Wales. In J. Katz, S. Peace, & S. Spurr (Eds.), *Adult Lives: A Life Course Perspective* (pp. 161–172). The Policy Press.

Goetting, A. (1984a). The Elderly in Prison: A Profile. *Criminal Justice Review, 9*(2), 14–24. https://doi.org/10.1177/073401688400900203.

Goetting, A. (1984b). Prison Programs and Facilities for Elderly Inmates. In E. S. Newman, D. J. Newman, & M. L. Gewirtz (Eds.), *Elderly Criminals* (pp. 169–175). Oelgeschlager, Gunn & Hain, Publishers, Inc.

Goetting, A. (1985). Racism, Sexism, and Ageism in the Prison Community. *Federal Probation, 44*(3), 10–22.

Golden, D. (1984). Elderly Offenders in Jail. In E. S. Newman, D. J. Newman, & M. L. Gewirtz (Eds.), *Elderly Criminals* (pp. 143–152). Oelgeschlager, Gunn & Hain, Publishers, Inc.

Grant, A. (1999). Elderly Inmates: Issues for Australia. *Trends and Issues in Crime and Criminal Justice, 115*, 1–5.

Greene, S. (2014). *The Psychological Development of Girls and Women: Rethinking Change in Time*. Routledge.

Greve, W., & Staudinger, U. M. (2006). Resilience in Later Adulthood and Old Age: Resources and Potentials for Successful Aging. In D. Cicchetti & D. J. Cohen (Eds.), *Developmental Psychopathology. Vol. III: Risk, Disorder, and Adaptation* (2nd ed., pp. 796–840). John Wiley & Sons.

Gubler, T. (2006). *Elderly Prisoners Are Literally Dying for Reform* (SSRN Scholarly Paper ID 977249). Social Science Research Network. Retrieved from http://papers.ssrn.com/abstract=977249.

Gubrium, J. F., & Wallace, J. B. (1990). Who Theorises Age? *Ageing & Society, 10*(2), 131–149. https://doi.org/10.1017/S0144686X00008047.

Hagestad, G. O. (2008). The Book-Ends: Emerging Perspectives on Children and Old People. In C. Saraceno (Ed.), *Families, Ageing and Social Policy: Intergenerational Solidarity in European Welfare States* (pp. 20–37). Edward Elgar Publishing.

Hagestad, G. O., & Herlofson, K. (2007). *Micro and Macro Perspectives on Intergenerational Relations and Transfers in Europe*. Department of Economic and Social Affairs, United Nations Organization (UNO). Retrieved from http://www.un.org/esa/population/meetings/Proceedings_EGM_Mex_2005/hagestad.pdf.

Harrison, M. T. (2006). True Grit: An Innovative Program for Elderly Inmates. *Corrections Today, 68*(7), 46–49.

Hayes, A. J., Burns, A., Turnbull, P., & Shaw, J. J. (2013). Social and Custodial NEEDS of Older Adults in Prison. *Age and Ageing*. https://doi.org/10.1093/ageing/aft066.

Hayslip, B., Jr., & Panek, P. E. (1993). *Adult Development and Aging*. Harper Row.

Hermsen, J. (2013). Tijd en verveling. *Podium voor Bio-ethiek, 20*(2), 3–7.

House of Commons Justice Committee. (2013). *Older Prisoners: Fifth Report of Session 2013–14. HC 89.* Report, together with formal minutes, oral and written evidenc; p. 144. The Stationery Office.

Howse, K. (2003). *Growing Old in Prison: A Scoping Study on Older Prisoners* (p. 49). Prison Reform Trust.

Human Rights Watch. (2012). *Old Behind Bars: The Aging Population in the United States*. Human Rights Watch. Retrieved from https://www.hrw.org/report/2012/01/27/old-behind-bars/aging-prison-population-united-states.

Humblet, D. (2014). *Imprisoned Older Adults: Patients or Prisoners?* (unpublished paper), 1–13.

Humblet, D., & Decorte, T. (2013). Detentiebeleving door oudere onderzoekgedetineerden in België: Een exploratief. *Panopticon. Tijdschrift voor strafrecht, criminologie en forensisch welzijnswerk, 34*(4), 267. Retrieved from http://hdl.handle.net/1854/LU-3238404.

Humblet, D., & Van Limberghen, G. (2017). Ouderdom en detentie: De ondraaglijke lichtheid van de 'bestaanszekerheid'? *Fatik. Tijdschrift voor Strafbeleid en Gevangeniswezen, 34*(154), 24–35.

Hutchison, E. D. (2014). *Dimensions of Human Behavior: The Changing Life Course*. SAGE Publications.

Ikuteyijo, L. O., & Rotimi, A. R. (2009). Elderly Prisoners and the Nigerian Criminal Justice System. *Indian Journal of Gerontology, 24*(4), 517–528.

International Longevity Center. (2006). *Ageism in America*. The Anti-Ageism Taskforce, International Longevity Center. Retrieved from http://aging.columbia.edu/sites/default/files/Ageism_in_America.pdf.

Jiang, Y., Jachna, T. J., & Dong, H. (2016). Understanding the Critical Needs of Older People: An Aging Perspective. In J. Zhou & G. Salvendy (Eds.),

*Human Aspects of IT for the Aged Population. Design for Aging: Second International Conference, ITAP 2016, Held as Part of HCI International 2016, Toronto, ON, Canada, July 17–22, 2016, Proceedings* (pp. 23–31). Springer International Publishing.

Jurkowski, E. T. (2008). *Policy and Program Planning for Older Adults: Realities and Visions*. Springer Publishing Company.

Kaufman, S. R. (1986). *The Ageless Self: Sources of Meaning in Late Life*. University of Wisconsin Press.

Kerbs, J. J., & Jolley, J. M. (2007). Inmate-on-Inmate Victimization Among Older Male Prisoners. *Crime & Delinquency, 53*(2), 187–218. https://doi.org/10.1177/0011128706294119.

Kerbs, J. J., & Jolley, J. M. (2009). A Commentary on Age Segregation for Older Prisoners Philosophical and Pragmatic Considerations for Correctional Systems. *Criminal Justice Review, 34*(1), 119–139. https://doi.org/10.1177/0734016808324245.

Kowal, P., & Dowd, J. E. (2001). *Definition of an Older Person. Proposed Working Definition of an Older Person in Africa for the MDS Project*. World Health Organization. Retrieved from http://www.who.int/healthinfo/survey/ageing-defnolder/en/.

Lemieux, C. M., Dyeson, T. B., & Castiglione, B. (2002). Revisiting the Literature on Prisoners Who Are Older: Are We Wiser? *The Prison Journal, 82*(4), 440–458. https://doi.org/10.1177/0032885502238680.

Liebling, A., & Maruna, S. (2011). Introduction: The Effects of Imprisonment Revisited. In A. Liebling & S. Maruna (Eds.), *The Effects of Imprisonment* (2nd ed., pp. 1–30). Routledge.

Loeb, S. J., Steffensmeier, D., & Lawrence, F. (2008). Comparing Incarcerated and Community-Dwelling Older Men's Health. *Western Journal of Nursing Research, 30*(2), 234–249. https://doi.org/10.1177/0193945907302981.

Maelstaf, H., & Claessens, D. (2016). *Oudere gedetineerden, onbekend en onbemind* (p. 169). Artesis Plantijn Hogeschool Antwerpen. Retrieved from http://www.kenniscentrumisos.be/wp-content/uploads/2017/06/Oudere-gedetineerden.pdf.

Mann, N. (2008). *Doing Harder Time? The Experience of an Ageing Male Prison Population in England and Wales*. University of Essex.

Mann, N. (2012). *Doing Harder Time? The Experiences of an Ageing Male Prison Population in England and Wales*. Ashgate.

Marquart, J. W., Merianos, D. E., & Doucet, G. (2000). The Health-Related Concerns of Older Prisoners: Implications for Policy. *Ageing & Society, 20*(1), 79–96.

Maruna, S. (2001). *Making Good: How Ex-convicts REFORM and Rebuild Their Lives* (Vol. xix). American Psychological Association.

Maschi, T., Viola, D., & Sun, F. (2012). The High Cost of the International Aging Prisoner Crisis: Well-being as the Common Denominator for Action. *Gerontologist, 53*(4), 543–554.

McAdams, D. P. (2013). *The Redemptive Self: Stories Americans Live by - Revised and Expanded Edition*. Oxford University Press.

McShane, M. D., & Williams, F. P. (1990). Old and Ornery: The Disciplinary Experiences of Elderly Prisoners. *International Journal of Offender Therapy and Comparative Criminology, 34*(3), 197–212. https://doi.org/10.1177/0306624X9003400304.

van der Meide, H., & Olthuis, G. (2013). "Mensen vervelen zich hier, hè?" Over oudere patiënten en verveling in het ziekenhuis. *Podium voor Bio-ethiek, 20*(2), 30–32.

Ministry of Justice. (2013). *Government Response to the Justice Committee's Fifth Report of Session 2013–14. Older Prisoners*. The Stationery Office Limited.

Moll, A. (2013). *Losing Track of Time: Dementia and the Ageing Prison Population: Treatment Challenges and Examples of Good Practice*. Retrieved from https://www.mentalhealth.org.uk/publications/losing-track-time

Moore, E. O. (1989). Prison Environments and Their Impact on Older Citizens. In S. Chaneles & C. Burnett (Eds.), *Older Offenders: Current Trends* (pp. 175–192). The Haworth Press, Inc.

Moran, K. (2014). *Generativity Paper*. Unpublished. Retrieved from http://crimworkshop.ws.gc.cuny.edu/files/2013/08/Generativity-Paper3.docx.

Morton, J. B. (1992). *An Administrative Overview of the Older Inmate*. National Institute of Corrections. Retrieved from http://www.ncjrs.gov/App/abstractdb/AbstractDBDetails.aspx?id=144056.

Morton, J. (2001). Implications for Corrections of an Aging Prison Population. *Corrections Management Quarterly, 5*(1), 78–88.

Neugarten, B., & Hagestad, G. (1976). Age and the Life Course. In R. Binstock & E. Shanas (Eds.), *Handbook of Aging and the Social Sciences* (pp. 35–55). Van Nostrand Reinhold.

Newman, E. S., Newman, D. J., & Gewirtz, M. L. (1984). Varieties of Elderly Crime. In *Elderly Criminals* (pp. 33–34). Oelgeschlager, Gunn & Hain, Publishers, Inc.

O'Brien, J. (2011). Epilogue. In J. O'Brien (Ed.), *The Production of Reality: Essays and Readings on Social Interaction* (5th ed., pp. 517–524). Pine Forge Press.

O'Donnell, I. (2014). *Prisoners, Solitude, and Time.* Oxford University Press.

Pachana, N. A. (2016). *Ageing: A Very Short Introduction.* Oxford University Press.

Pain, R. (1999). Theorising Age in Criminology: The Case of Home Abuse. In M. Brogden (Ed.), *British Criminology Conferences* (pp. 1–12). British Criminology Society. Retrieved from http://www.britsoccrim.org/volume2/006.pdf.

Palmore, E. (1999). *Ageism: Negative and Positive* (2nd ed.). Springer Publishing Company.

Peterson, C., Maier, S. F., & Seligman, M. E. P. (1993). *Learned Helplessness: A Theory for the Age of Personal Control.* Oxford University Press.

Potter, E., Cashin, A., Chenoweth, L., & Jeon, Y.-H. (2007). The Healthcare of Older Inmates in the Correctional Setting. *International Journal of Prisoner Health, 3*(3), 204–213.

Prevc, P., & Doupona Topic, M. (2009). Age Identity, Social Influence and Socialization Through Physical Activity in Elderly People Living in a Nursing Home. *Collegium Antropologicum, 33*(4), 1107–1114.

Reed, M. B. (1978). *Aging in a Total Institution: The Case of Older Prisoners* (pp. 1–84). Tennessee Corrections Institute. Retrieved from https://www.ncjrs.gov/pdffiles1/Digitization/65232NCJRS.pdf.

Roberts, S. K. (2015). *Aging in Prison: Reducing Elder Incarceration and Promoting Public Safety.* Center for Justice at Columbia University.

Rozario, P. A., & Derienzis, D. (2009). 'So Forget How Old I Am!' Examining Age Identities in the Face of Chronic Conditions. *Sociology of Health & Illness, 31*(4), 540–553. https://doi.org/10.1111/j.1467-9566.2008.01149.x.

Rubenstein, D. (1982). The Older Person in Prison. *Archives of Gerontology and Geriatrics, 1*, 287–296.

Sapp, A. D. (1989). Arrests for Major Crimes: Trends and Patterns for Elderly Offenders. *Journal of Offender Counseling, Services & Rehabilitation, 13*(2), 19–44.

Schroots, J. J. F., & Birren, J. E. (2013). Concepts of Time and Aging in Science. In J. E. Birren & K. W. Schaie (Eds.), *Handbook of the Psychology of Aging* (3rd ed., pp. 45–66). Academic Press Inc.

Settersten, R. A., & Mayer, K. U. (1997). The Measurement of Age, Age Structuring, and the Life Course. *Annual Review of Sociology, 23*, 233–261.

Settersten, R. S., & Mayer, K. U. (2002). The Measurement of Age, Age Structuring, and the Life Course. In D. L. Infeld (Ed.), *Disciplinary Approaches to Aging* (pp. 323–352). Routledge.

Snacken, S. (2001). Van ongelijke levensvoorwaarden tot geweld in gevangenissen: Het belang van gevangenisregimes. In L. Dupont & F. Hutsebaut (Eds.), *Herstelrecht tussen toekomst en verleden* (pp. 455–478). Universitaire Pers Leuven.

Spaulding, A. C., Seals, R. M., McCallum, V. A., Perez, S. D., Brzozowski, A. K., & Steenland, N. K. (2011). Prisoner Survival Inside and Outside of the Institution: Implications for Health-Care Planning. *American Journal of Epidemiology, 173*(5), 479–487. https://doi.org/10.1093/aje/kwq422.

Stans, S. E. A., Dalemans, R., de Witte, L., & Beurskens, A. (2013). Challenges in the communication between 'communication vulnerable' people and their social environment: An exploratory qualitative study. *Patient Education and Counseling, 92*(3), 302–312.

Sterns, H. L., & McQuown, C. (2018). Aging, Work, and Retirement. In B. Yochin & E. Woodhead (Eds.), *Psychology of Aging: A Biopsychosocial Perspective*. Springer Publishing Company.

Stoller, S. (2014). The Coming of Age. In *Simone de Beauvoir's Philosophy of Age: Gender, Ethics, and Time* (pp. 1–28). Walter de Gruyter.

Sugarman, L. (2004). *Counselling and the Life Course*. London: SAGE Publications.

Sundberg, N. D. (2004). Boredom and Boredom Proneness. In W. E. Craighead & C. B. Nemeroff (Eds.), *The Concise Corsini Encyclopedia of Psychology and Behavioral Science* (pp. 133–135). John Wiley & Sons.

Svendsen, L. (2005). *A Philosophy of Boredom*. Reaktion Books.

van Turenhout, M., & van der Heijden, T. (2009). *Seniorencriminaliteit. Een verkennend onderzoek* (p. 160). Openbaar Ministerie.

United Nations Office on Drugs and Crime. (2009). *Handbook on Prisoners with Special Needs* (Criminal Justice Handbook Series). United Nations Publications.

Uzoaba, J. H. E. (1998). *Managing Older Offenders: Where Do We Stand?* Research Branch Correctional Service of Canada. Retrieved from http://www.csc-scc.gc.ca/research/092/r70_e.pdf.

Vanderbeck, R. M. (2007). Intergenerational Geographies: Age Relations, Segregation and Re-engagements. *Geography Compass, 1*(2), 200–221. https://doi.org/10.1111/j.1749-8198.2007.00012.x.

Wahidin, A. (2000). Life Behind the Shadows: Women's Experiences of Prison in Later Life. In R. Horn & S. Warner (Eds.), *Issues in Forensic Psychology: Positive Directions for Women in Secure Environments* (pp. 53–58). British Psychological Society.

Wahidin, A. (2004). *Older Women in the Criminal Justice System: Running Out of Time*. Jessica Kingsley Publishers.

Wahidin, A. (2005). We Are a Significant Minority: Elderly Women in English Prisons. *British Journal of Criminology*. Retrieved from http://www.britsoc-crim.org/volume6/001.pdf.

Wahidin, A., & Aday, R. H. (2005). The Needs of Older Men and Women in the Criminal Justice System: An International Perspective. *Prison Service Journal, 160*, 13–22.

Wangmo, T., Handtke, V., Bretschneider, W., & Elger, B. S. (2017). Prisons Should Mirror Society: The Debate on Age-Segregated Housing for Older Prisoners. *Ageing & Society, 37*, 675–694.

Weiss, R. S., & Bass, S. A. (2002). Introduction. In R. S. S. Weiss & S. A. Bass (Eds.), *Challenges of the Third Age: Meaning and Purpose in Later Life* (pp. 3–12). Oxford University Press.

Westerhof, G. J. (2008). Age Identity. In D. Carr (Ed.), *Encyclopedia of the Life Course and Human Development* (Vol. 3, pp. 10–14). Macmillan. Retrieved from http://midus.wisc.edu/findings/pdfs/798.pdf.

Whitbourne, S. K. (2012). What's Your True Age? You May Be a Lot Younger than You Think. *Psychology Today*. Retrieved from http://www.psychologytoday.com/blog/fulfillment-any-age/201206/what-s-your-true-age.

Williams, B. A., & Abraldes, R. (2007). Growing Older: Challenges of Prison and Reentry for the Aging Population. In R. Greifinger (Ed.), *Public Health Behind Bars: From Prisons to Communities* (pp. 56–72). New York: Springer.

Williams, B. A., Lindquist, K., Sudore, R. L., Strupp, H. M., Willmott, D. J., & Walter, L. C. (2006). Being Old and Doing Time: Functional Impairment and Adverse Experiences of Geriatric Female Prisoners. *Journal of the American Geriatrics Society, 54*(4), 702–707. https://doi.org/10.1111/j.1532-5415.2006.00662.x.

Williams, B. A., Stern, M. F., Mellow, J., Safer, M., & Greifinger, R. B. (2012). Aging in Correctional Custody: Setting a Policy Agenda for Older Prisoner Health Care. *American Journal of Public Health, 102*(8), 1475–1481. https://doi.org/10.2105/AJPH.2012.300704.

Woodward, K. M. (1991). *Aging and Its Discontents: Freud and Other Fictions*. Indiana University Press.

Youdin, R. (2014). *Clinical Gerontological Social Work Practice*. Springer Publishing Company.

Zevallos, Z. (2011, oktober 14). What Is Otherness? *The Other Sociologist*. Retrieved from https://othersociologist.com/otherness-resources/.

Zoutendijk, A. (2007). *Bejaarde Boeven. Bejaarde gedetineerden in Nederland en hun detentiebeleving*. Scriptie Master Criminologie. Universiteit Utrecht.

Zoutendijk, A. (2009). Bejaarde gedetineerden in Nederland. *PROCES, 88*(5), 243–262.

# 3

# The Older Prisoner and the Physical Prison World

It has been stated that prison space, which is predominantly characterised by high and thick walls, reinforced doors, barbed wire fences, and tight security, is there to demonstrate that 'physical and social exclusion is the price of non-conformity' (Matthews 1999: 26; Mann 2013: 11). Bearing in mind the double non-conformity of being an older prisoner, as discussed in Chap. 2, what do these prison environments bring about for them? The dilemmas and contradictions which surround the notion of housing adults in(to) old age in such environments are perhaps best documented by the following statement:

> I don't want to feel sympathetic towards these people [ageing prisoners] … what evil do you have to do to get 30 years? … But why are there bars on the window of a man who can't walk without a frame? What kind of escape plan can be hatched by a man who can't remember how to go to the toilet? (Citation of S. Norfolk in the foreword to the photographic book of Clark (2008)[1] and referred to by Mann (2013: 11, own emphasis))

Given that the prison environment differs markedly to that in which many individuals in wider society grow old,[2] it is unsurprising that the ageing prison population experience difficulties in such antiquated



D. Humblet, *The Older Prisoner*, Palgrave Studies in Prisons and Penology,
https://doi.org/10.1007/978-3-030-60120-1_3

institutions that are typically designed to hold and control young men (Mann 2013: 11; Ginn 2012: 24; Howse 2003: 1). An emerging body of research on the ageing prisoner population (Mann 2012; Crawley and Sparks 2005a, b; Wahidin 2000, 2004, 2005) has concluded that older prisoners are punished even further by the prison environment (Mann 2013: 11).

Crawley and Sparks pointed to the most obvious aspects in which prisons have never been designed with older people and their needs in mind:

> Their very fabric (the stairs and steps and walkways, the distances, the gates, the football pitches and gymnasia; the serveries and queues; the communal showers; the incessant background noise) is, in general, constructed in blithe unconsciousness of the needs and sensibilities of the old. (Crawley and Sparks 2005a: 350, own emphasis)

Prison epitomises the idea of constraint and thus inherently restricts the actions of the individuals which it seeks to punish (Mann 2012: 42).[3] However, for an ageing prisoner population, the constraints of the prison building become even more challenging—in some cases even damaging (Mann 2013: 13). Glover (2012: 163) puts forward the idea that the architects of prisons from Victorian times onwards apparently assumed that prisoners would be young and able-bodied and gave no further consideration to those who, although apparently healthy on arrival, may become ill or disabled during their incarceration. This thoughtlessness is often reflected in the design of prison buildings, many of which are characterised by, inter alia, small cells, no purpose-built accommodation for older and disabled people, narrow doorways and corridors, and several storeys reached by steep, open staircases (Glover 2012: 163). Many prisoners who are frail, or are functionally impaired, find that they cannot cope with the constant noise or are unable to access some parts of the prison (Glover 2012: 163).

In Belgium, as in many other countries, most older prisoners are not housed in specialist penitentiary establishments. The two field sites thus provide an interesting juxtaposition of a common 'integrated' variant (prison setting no. 1) with a rare 'segregated' variant (prison setting no. 2), to understand and assess the phenomenon of growing older within the prison realm. The prison environment is approached here both as a mathematical space, one which is 'measured in three dimensions, in metres and centimetres' (Rommel et al. 1998: 17; Bollnow 1961, 1963), and as a 'lived-space', i.e. 'space itself and, in so far as humans live in it and with it, to space as medium of human life' (Bollnow 1963: 19).

## Mathematical Space

### The First Prison Setting[4]

The first site comprises one of the largest prisons in Belgium since 1991 and accommodates both remanded and sentenced prisoners. The penitentiary complex has a department for sentenced prisoners serving sentences of more than five years (termed 'Men 1') and a section for both prisoners on remand and prisoners sentenced to a sentence of less than five years ('Men 2'), as well as a department for women ('Women'). The establishment also has a psychiatric annex for female offenders.

Most Belgian prisons are of the 'Ducpétiaux' type, whereby the cell wings are shaped as a star and are implanted around a central observation post. Moreover, each cell wing has multiple floors construed as galleries. Ducpétiaux advocated the total seclusion of prisoners in individual cells. Prison setting no. 1 was established in the context of a—at that time (1970s–1980s)—new philosophy: i.e. the inmates are divided into small communities, with, however, several more centrally situated and shared accommodations.

This penitentiary complex opened in 1991 and consists of one vast building conceived as three independent establishments with a common structure. It is one of the largest prisons in Belgium, with a capacity of 512 male and 114 female remand and sentenced prisoners. It houses a psychiatric wing for mentally ill offenders, a medical centre with a hospital, polyclinic, and pharmacy, a day-leave section, and a high-security section. Prison setting no. 1 occupies a central place within the Flemish prison system, as it provides services to several other prisons in Flanders:

amongst which, inter alia, a medical centre, an industrial laundry, and more than 7.000 $m^2$ of industrial workshops and warehouses.

Although conceived as a 'normal' closed prison, the architectural design and technological security have turned it into a maximum-security institution. Flat outside walls in yellow-pink masonry and continuous lines of black windows replace the traditional prison bars with reflective and sunproof security glass which is surrounded by a concrete fencing wall with a cylindrical head and located in a three-metre-deep moat. The entire complex has remotely controlled doors, fences, and gates, and the surveillance system includes central control, electrical locks, interphone and telephone system, CCTV monitoring, prison perimeter detection, night-shift control, access control, metal detection (metal detection porticos are installed along the circulatory pathways from outside to inside or from the workplaces and visitor spaces to the cellular departments), operating panels, and synoptics.

It was built according to contemporary notions of what a prison ought to look like. To that end, it was designed to organise prisoners into small communities of 20–30 prisoners, who make use of the common facilities situated in the centre of the complex. The various departments are grouped around a large central block. All units arrive at two perpendicular central corridors, without crossing of different groups and functions. The prison is cross-shaped with honeycomb-like arms constituting each cell block of two floors for 21 persons. The choice of a cross shape with comb-type arms should facilitate working with small groups of prisoners. Each arm represents a cell block. Each block is composed of 17 individual cells and one group cell for four prisoners. For each group there is a multipurpose hall where prisoners can eat and drink collectively. There is also a shower facility. In practice, the communal living is limited to a common moment in the evening when the inmates may reside in the multipurpose area on their section, according to the over-population ('polyvalent room') (Annual report of the prison: De Jong 2012), each department also has a classification ('klassement') with a number of agencies and clothing depot ('Bath') on the first floor. On the second floor of the executive agencies and call areas of the PSD, and JWW, are other individual service providers.

Each cell is allotted approximately ten square metres. The interior has two windows, a steel outer frame and a wooden inner frame connected with bolts, although a smaller part of the window can be opened. The space of the cell is dominated by neutral beige and white, and the sliding heavy metal doors let in virtually no light save for the small observation hatches at the eye level. Each cell has a wooden wall table, a wooden or plastic chair, a wooden bed, and a cabinet. There is a separate space with a washbasin, toilet, and mirror and connections for radio and TV. With the exception of floor heating in high-security cells, the wings and cells are heated with steel radiators. Each cell is equipped with a push-button intercom, a lamp with two-colour fields, and a push-button outside the cell that can be turned on by prison officers at night to perform a visual check.

The construction of the building has been predominantly guided by a rationalisation and safety imperative: the clustering of the various departments around a large central block is both economically sustainable and functionally viable. All units open on two central corridors perpendicular to each other to enable independent crossing; positioning oneself within which enables the circulation of prisoner traffic to be easily and safely observed.

Each arm represents a department, with a number of offices and a clothes depot ('Bath') on the first floor. The offices of the prison managers, the psycho-social services (PSS), judicial welfare workers (JWW), and other individual service providers are found on the second floor. The common infrastructure such as the sports centre and gym, theatre room, and places for worship are centrally located. The visitors' rooms are located on the first floor. One level down, there is the practice room, the library and the educational centre. On the ground floor, there is the central prison kitchen and a cafeteria for prison staff.

This prison could be considered as the epitome of a highly enforced closed prison (Beyens and Maes 2017: 262f.), with a prevailing closed-door prison regime inside the various units. Prisoners are mostly supervised from secure control booths at key points and the central supervisory office (invisible CCTV), whilst prison officers have little or no contact with inmates. Whilst the architectural style of the building is aimed at small groups of inmates working together, the life group functioning is, in practice, limited to one single common moment in the evening. General messages about directions, work, and alarms are communicated via the intercom, whilst more personal interactions take place through the prison door hatch, which was justified by prison officers with 'If it isn't necessary, we're not going to open the door'.

This prison integrates its older inmates amongst the rest of the prisoner population. Older prisoners are here subjected to the same prison regime as their younger counterparts. At the time of the data collection, most older prisoners (65+) were housed in mono cells. Most core participants included in the study resided at the department for long-term sentenced prisoners.

## The Second Prison Setting[5]

The prison was built in 1825 as a colony to take in beggars and vagrants. After World War II, it evolved into a semi-open prison regime community. To date, it is a closed establishment for male prisoners under sentence. Built in 1825 as a workhouse for vagrants, this prison complex is located in a unique historic landscape and was classified as a cultural-historical monument in 1999, imposing severe restrictions on any architectural change or renovations.

It is part of a vast grid-shaped domain of lanes, fields, forests, meadows, and quarries, situated in a remote location in the north of the country. After growing up in this region, the author Horsten wondered if it was 'a coincidence that the Belgian society locked its outcasts away on the edge of the country' (Horsten 2013).

The prison complex is subdivided into ten pavilions covering a large site of 66.25 hectare. A characteristic feature of prison setting no. 2 is the breakdown of the cellular complex in eight pavilions. Only one pavilion has standard individual cells. The other pavilions accommodate most prisoners in multiple cells (duos, quatros, and some larger living and sleeping rooms). The accommodation is largely composed of multiple occupation cells where prisoners are clustered into dual, quadruple, and some larger dormitories and communal living quarters. Single cells are only available in two pavilions. The overall capacity has gradually declined because of closures and the more recent destruction of two pavilions after a prisoner uprising against overcrowding and poor prison conditions. The total number of inmates is now 366.

This prison employs CCTV surveillance and fencing to secure its external perimeter. Over the last years, the prison evolved from a semi-open prison to one more closed, albeit with an open-door prison regime inside (Beyens and Maes 2017: 262). This institution arguably combines its function of social protection with that of punishment (Vanden Hende 2005: 28).

The main building opens onto a large outside cobblestoned area with a lane of trees and an obsolete fountain, which must be crossed to reach Pavilion V, which houses the 'gero-medical unit', where the fieldwork took place. The entrance is enclosed by a large metal gate which must be opened manually. The inside of Pavilion V can only be accessed via stairs or a steep ramp.

The special unit, i.e. the second field site, is informally described by staff as 'the old men's home' or the 'hospital' (after its former function as the hospital for vagrants). Only a minority of ageing inmates live in this dormitory-style unit, together with a few younger disabled inmates, the others living amongst the general prison population. Upon entry there were approximately 54 people housed in Pavilion V, 13 of whom resided in the Ward 1, the unit under study.

The Mental Health Foundation report Losing Track of Time (Moll 2013) notes that this penitentiary institution (hereafter referred to as prison (setting) no. 2) 'houses its oldest and most infirm inmates on a separate wing, although the building's listed status prevents radical overhauls and has determined that the layout hasn't changed a great deal since it was opened as a homeless colony in 1825'. From the report of Moll (2013), it follows that criteria for admission to 'Ward 1', the specialist unit for old or infirm prisoners, include good behaviour as it is not deemed safe to manage violent offenders of any age. Specialist accommodation for older prisoners is a rarity and a luxury not afforded by most facilities (cf. prison setting no. 1). Although this specialist unit de facto houses a majority of older prisoners (not always +65), it is a unit for prisoners who require extra (medical) care regardless of age. Moreover, it must be stressed that older prisoners are also housed in other units of the prison, integrated within the rest of the prisoner population.

Ward 1 is an L-shaped space with high windows that looks out onto a paved courtyard or onto the hallway. With a total capacity for 20 prisoners, it is composed of (1) a dormitory accommodating up to 14 persons, (2) four single cells, (3) one double cell (reserved for the two prisoner-servants), and (4) one large communal living quarters. In addition, there are two toilets and one toilet for disabled people, a scullery, a smoking room, a room with showers, and one pivoting bathtub. The four separate rooms are allocated to those who require extra intensive care. The prisoners (except for the prisoner servants) sleep in hospital beds. Where the older inhabitants sleep in a dormitory the issue of privacy was raised. In the dayroom, there is an aquarium tank with fish, a closet with games, a television, and several sofas, seats, and chairs.

In contrast to the other prison wings and units, Ward 1 requires much less opening and closing doors with keys as most prisoners are housed in a dormitory and can freely move around. The inhabitants of this unit are subject to adapted, more open, regime. For instance, they are allowed to walk around freely from the sleeping area to the dayroom (even at night). Conversely, the rooms of the mono cells are locked at night. There is a psychiatric nurse who provides daily assistance to the prisoners and carries out activities with them. The healthcare lead at prison setting no. 2, where prisoners are assessed on a case-by-case basis, expressed a wish to be able to conduct more regular checks but stated staff capacity prevented this. Prison staff also stressed the importance of older offenders, when mentally capable, being able to take charge of their health and remain independent.

Aside basic elements like security grills and fences, there are relatively few indications of passive security. The interactions between staff and prisoners are primarily guided by the principles of dynamic security with greater levels of interaction in comparison to the first prison setting. However, towards the end of the fieldwork, some new elements of passive safety were being implemented in the unit: bars in front of the windows and an additional metal-grilled gate in the hallway.

Prison officers referred to the poor passive security ('bad doors, single glass, cardboard doors') yet did not express a heightened sense of insecurity as a result. They were more concerned to make arguments in favour of investment in the facilities for prisoners, such as enhancing the gero-medical unit or repairing the showers, than in upgrading the passive security. Although this may also reflect their own preference not to be subjected to observation: 'we can handle it ourselves just fine', they said scornfully.

In contrast to prison setting no. 1, direct supervision is rooted in the physical organisation of space. Whilst the dormitory and private rooms are bounded by doors, the corridor connecting the dormitory with the common living area is not enclosed. As a result, prison officers are stationed inside the prisoners living area. Supervision is a matter for various actors: prison officers, nurses, psychiatric nurses, and prisoners. However, as prisoners in the gero-medical unit represent a low-security threat, prison officers would only sporadically pop in. This was even the main reason for some prison officers to select this unit:

> Yeah, but I prefer those old prisons, like this one'. I don't want anything to do with those cameras and that control and such … Over here, no one else can see what's going on'. […] I ask Joe why he chose to work there. He says: 'The main reason, is actually because you're often working alone in here … And then you can do what you want to, right?'. And I also want to take it a little slower. At the end of the day, you get equal pay, regardless of whether you work hard or not. […] (Joe)

This prison setting also utilises volunteers, from JWT—a judicial welfare group, who attends the establishment every two weeks to offer relaxing activities designed for older inmates including walking and home cycling. Moreover, the prison has a (psychiatric) nurse who plays specialised games with older prisoners designed to benefit those with memory problems. They use their volunteers to assist with exercises on the wings, and staff have reported that this has had a very positive effect on the inhabitants. Moreover, they have tasks such as cleaning the paths in the yards, feeding their fish and watering the plants, although the prison reports that in reality little gets done.

## The Lived Space

Lived space refers to how the objective mathematical space is *experienced* (Bollnow 1961), which has been uncovered by the core participants for both field sites.

### Prison Setting No. 1

From the prisoner's most inner point of perception, the prison environment was considered as a concentric structure, thus beginning our description at their cell as the starting point and branching out from there to the outermost parts of the circle, from the living group to the common areas (Rommel et al. 1998: 35).

#### Privacy: Control

The need for privacy was widely acknowledged by the participants and contributed to the significance of the in-cell environment. For a small minority of older prisoners, the prison cell was merely a place to sleep, whereas for an overwhelming majority, it represented a critical site of multiple functions. All core participants resided on their own in a single cell, which allowed for some personal space, intimacy, and withdrawal (Milhaud and Moran 2016), which was highlighted in many interviews. For example, both Dorothy and Norman cited the ability to return to one's cell was very useful if they were overwhelmed by emotions as a result of the actions of prison staff. Fellow prisoners were not allowed to stay in a prison cell that was not allocated to them, and prison officers were not allowed to enter the room of a prisoner alone, although this occasionally happened.

The ten square metres left little surface area for 'ordinary' circulation space within cells. This was exacerbated for prisoners with high functional ages, wheelchair, or rollator users and for those with reduced mobility, for whom it poses a challenge to manoeuvre oneself in, within, and out of the cell. As a result, some older prisoners were confined to their cell. Moreover, falling was a fairly common risk factor amongst these prisoners, raising the dilemma between the isolation and absence of continuous observation in case of necessity versus the possibility for prison officers to invade their personal space without their or prior notice. The cumbersome procedure to open a cell door at night, which required permission by the prison governor, was denounced by prisoners in both settings. Whilst this was introduced as a protection of prisoners against unnecessary intrusions by staff, it was felt to be potentially life-threatening in case of emergency.

The prison door hatch or 'peephole'—a residual panoptical element in the design of the prison—allows prison staff at any time 'to see without being seen' (Hoffman 2014; Jacobs and Heracleous 2001). Most participants regarded this feature as disruptive, with a psychological, and by proxy occasionally physical impact on the 'seen'. Whilst prison staff justified the use of the hatch on grounds of security and safety, this was questioned by the prisoners referring to the frequent prison staff strikes during which there appeared to be virtually no supervision. Being locked up under the supervision of a 'gaze' (Lacan in Hovey 2006: 27) hampered older prisoners' self-care:

> R: And otherwise, I might still do some exercises in my cell. No, I should actually do it, but I don't get around to it.
>
> I: How come you haven't got around to it?
>
> R: Because I think they're going to have a look at that little window while I'm doing exercises. And, well yes … that I'm going to feel ridiculous with what I'm doing. That's why I won't.
>
> I: I understand. Do they always see you?
>
> R: Well yes, huh… Not if you're sitting on the toilet. They cannot behind…
>
> I: Can't you do your exercises over there?
>
> R: Hahaha. That's a bit limited in space. (Billy)

Some places in the cell (e.g. the top shelves of the closet) were difficult for older prisoners to reach. Dorothy recalls how she was startled by a prison officer looking through the prison door hatch whilst standing on the chair grabbing something.

Prisoners could keep prison-approved televisions and radios. Many of those in the sample left the radio or television on—sometimes on mute—so as to have motion and/or sound. This contributed to feelings of control and the creation of a comforting environment whereby one could disengage from the surroundings. Leon, an older prisoner with multimorbidity, stressed how a relocation to the rooms at the medical centre of the prison would be subordinate to maintaining this 'privileged space'.

The prison environment can be noisy, and the prisoners occupied spaces close to the landings and adjacent prison cells. Whilst many older prisoners often complained about the noise stemming from younger prisoners' more boisterous behaviour, Oscar had to purchase a set of headphones because he suffered from hearing loss and turned up the sound of his television too loudly. Other disturbances interrupted their sleep, including gulls swarming around the building, prison officers dropping the lid of the peephole, and the light going on during the night checks. Prisoners attempted to gain some control over their direct environment by attaching something to the peephole to dampen any noise, lying under blankets, lying with their face down, or lying backwards to obstruct visibility.

In some cases, prison staff used the location of the inmate instrumentally. John, for example, an older less mobile diabetic prisoner, was relocated closer to the infirmary in order to reduce walking distances whilst simultaneously imposing a minimum level of physical exercise to get to the infirmary.

## The Experience of Lived Space in Older Prisoners

Despite being able to retreat into their own cell, prisoners sharply felt the lack of control over other spatial aspects:

> I: And do you think that being incarcerated here has an impact on your health?
>
> R: Yes. [...] Secondly, the abnormal circumstances in which you find yourself. Meaning, erm, that you're locked up here. The psychological effect of not having a door handle on the interior side of the door. Err, I

have conjugal visits, not for err having sexual relationships, but to talk. To talk, for two hours. Well, and over there you had a door handle on the … inner door. Peculiar, huh? Peculiar. (Ivan)

If cell doors were temporarily left open with little supervision, there was a perceived risk of theft or victimisation from other inmates. Prison officers do not knock, nor do they wait for permission to enter cells, which risks knocking over a prisoner (Dorothy).

Whilst it was not the dominant view amongst prison officers, one female officer was very aware of the lack of control experienced by prisoners. For this reason, she would avoid locking the door behind prisoners immediately after their return, to prevent the feeling that they are promptly locked up. In the same vein, some older prisoners spoke in favour of an open-door regime, at least within the prison unit. As Ivan stated, 'it is not said that we would leave our cell, but at least we would have the feeling that we would be able to'.

Control is considered as an important dimension in constructing a sense of belonging ('feeling home') (Van Dijk 2012), but none of our respondents felt 'at home' in prison:

I: And do you feel at home?

R: Home is home, huh. This is my dig … ['kot' in Dutch].

I: But do you try to make it as comfortable as possible?

R: I'm studying, so this is my student dig [laughs].

I: Ah, that's how you see it?

R: That's how I see it, yes. So … No, home stays home. Yes, yes, yes. [silent pause]. Yes, you try to make it work. But if you enter and they've checked your room… Literally everything is on the floor. The bedding is taken off, and so on. And then I think: 'Dear god'. Yeah, they always have to… they try to psychologically wear you down… […] But really feel at home over here, you can never feel that. (Dorothy)

Billy similarly commented:

The other day, it must have been a day or two or three ago, that a surveillant [prison officer], said to me when I came back from my walk: 'Alright, you can go back to your little home now'. I said: 'What?' 'My little home?' 'I'm never ever gonna consider this as my home!'.

The atmosphere and the characteristics of the building were generally referred to as a 'depressing' and 'disheartening':

> It's depressing, eh … In fact, it's sad. Take for instance, you also have you to go through that metal detector. That's traumatising in itself, right? But we have to go through that every single day, well, at least if you're outside the cell. Because if you go to the prison workshop, you get subjected to a search by a guard. (Ivan)

The high levels of security were also denounced by many participants:

> In my case, they have to perform detection by using a metal hand-held detector. That's some sort of device. Because I have an osteosynthesis of the thorax, well actually of the sternum. And erm when I worked in one of the prison workshops, everybody was allowed to leave except for me. I was the very last, really the 'terrorist' in fact. And then I was subjected to that detection, but I had to wait until everybody had left and then I was allowed to move forward. (Ivan)

## Physical Demands and Comfort

The cells for ageing prisoners were largely similar. The furniture and interior fixture used in the prison cells were drab and uniform. The durable, cost-efficient, and generally uncomfortable materials used seemed to embody the idea that prisoners themselves are of little value. Virtually all participants expressed dissatisfaction with the condition of prison beds and mattresses. This was not entirely surprising in view of the high prevalence of physical, particularly back, problems amongst them. This was acknowledged by the prison doctor. In the same vein, the low non-adjustable beds without handles raised challenges for some prisoners in terms of simply getting up and out. The remaining furniture was hard and offered little support for arms or feet and was not very ergonomic nor age-friendly, despite being often used for long periods at a time. Although adjustments were scarce, a refrigerator had been placed on the table for one frail prisoner, so it would be easier for him to use.

The limited windows in the building generally, and in the cells in particular, resulted in inadequate levels of natural light. Although natural light is important for any individual (Rommel et al. 1998: 79), it may affect older people even more (Sardegna et al. 2002: 120). This prompted Dorothy to request another cell:

> R: And then I went to a mono prison cell at the very front. But it's very dark at the very front. And then I requested another one… Well, since it's important with my work and such (e.g. drawing) that I can see clearly. And then I ended up in this one.
>
> I: Was that on the unit at the opposite side?
>
> R: No, I've always resided at this unit.
>
> I: Is there so much difference in the incidence of light?
>
> R: Yeah, yeah. In those cells at that side, they are very dark, because you're surrounded by four walls

And:

> I: Do you feel an effect of the physical environment?
>
> R: Hm, no, no. Not at all. Except for my eyesight that has worsened a lot. That's because you're continuously subjected to artificial lighting over here. (Dorothy)

Overall, the common attitude amongst our participants seemed resignation: 'it's prison; that's just the way it is'. It illustrates the risk mentioned by other authors that 'older' individuals receive less attention as they tend to make fewer demands on the basis that they have resigned themselves to existing structures (Baars 2012). However, the mere absence of expressed demands cannot be simply inferred that the interests and humanity of these older prisoners could not be undermined:

> I: What would you like to change about the infrastructure?
>
> R: Mattresses and a sealed cabinet. And for the rest … Well, what more could you want? You're in the gaol, so …. (Dorothy)

> Older prisoners who had previous experiences with prison compared their current environment positively with the former 'primitive' prison conditions:
>
> R: Its's nothing like it was thirty years ago. I still remember entering there and… But that was still in the *Pandreitje*, back in those days, the cells were quite different from the cells of today, pal. Gosh!
>
> I: What kind of cells were there?
>
> R: Well, gosh! Here, you have television. The only thing that you had over there was a box with 'on/off' above your cell and that was your radio. And it was all. That was everything. (Interview Billy, November 2014)

Leroy equally took the view that some of the younger prisoners complained too much: 'you have everything over here', referring to sport, television, computer, and so on. This view seemed to be particularly adopted by older prisoners who had experienced austere institutions in their own younger life, such as boarding schools, aboard ships, military service, or prisons.

## Individuality: Identity and Connectedness

Through the physical environment prisoners have the potential to represent themselves and their lives, through inter alia: pictures, artwork, writing letters. Most older prisoners in this setting had opportunities in this regard, as they all occupied single cells. Whilst there was some space to display personal effects, this was often restricted by the prison for reasons of order and liveability in the unit. For example, one of the participants was asked to remove a piece of artwork, a sculpture that had been created with food products, due to the smell that penetrated the lives of other prisoners and staff.

An inherent feature of the prison environment is the fetish for standardisation and the stripping away of individuality and anything which dissipates 'homeliness' and communicates the principle of uniformity. However, as noted above, neutrality equated to a less homely and a more dull, uninviting, and alienating environment. Prisoners were restricted in the things they can purchase, and they were held accountable for prison property. Eugene mentioned how he had once had to pay for a chair, broken as a result of falling backwards whilst falling asleep. He also stated that he barely escaped from having received a disciplinary report due to the accidental nature of the event.

Dale (a 60-year-old prisoner-servant who became a participant during the field research) was described by prison officers as 'institutionalised' and was perceived as part of the prison infrastructure. Although he was about to be released from prison after serving a sentence of 20 years, he strongly associated with the environment and delineated his identity (partly) from what others said he was:

> A little later, Dale also says that prison staff say the following about him: 'We're going to miss that old furniture over here'. I ask him why they say that, whereupon he replies: 'Well, because I'm old'. I ask: 'Why do you say furniture?'. He answers: 'Because I'm part of the prison'.

Alongside the projections from officers, or other prisoners, the physical environment of a penal institution might also play a role in shaping the identities of its inhabitants. Prisoners might have few or weak connections to their pre-prison environment yet still generally felt disconnected from the prison environment. This was especially the case when the prison environment was experienced as noisy or unsafe. Other prisoners however felt cut-off from their families and their familiar environments in later life. The cells located on the second or third floor served—in combination with the prison design and remote location—as an additional barrier to connect with the rest of the environment. In this context, life outside the institution, general society, and contact with nature to a considerable extent has dissolved.

In addition, a large amount of prison cells overlooked other cell blocks; there were very limited opportunities for prisoners to feel connected with the seasons. There was however enough space outside to create a proper (vegetable) garden, which was suggested by one of the participants.

In sum, there was little sense in which 'making prisoners feel at home' in terms of withdrawal or retreat, control, connectedness, and identity (Van Dijk 2012; Gifford et al. 2011) manifested itself in any meaningful way.

## Out-Cell Accessibility

Everything takes place within the same building, reflecting a micro society 'heterotopia'. Once prisoners left their individual cell—or by extension the prison unit—they were in unfamiliar 'alien territory'. The physical prison infrastructure allows at best for a degree of qualified independence and, at worst, for truly little, if any.

The 'communal' facilities, or what Rommel et al. (1998) refer to as the secondary and tertiary environments, were also located within the institution (e.g. chapel, recreation, visitation rooms, infirmary, workshops). The education centre was decentralised which inhibits movements to and from it (De Jong 2012: 6, 8). The telephones were difficult for wheelchair-bound prisoners to reach and awkward for those with cognitive problems. Some prisoners at high risk of falling over resorted to hanging on to the telephone equipment to prevent falling over.

With no own showers in cells, some were reluctant to shower for privacy, hygienic, and medical reasons or for feelings of insecurity regarding both other prisoners and/or the risk of falling. Whilst there were handy grab bars in some showers and toilet rooms, these were the exception rather than the rule. The request from officers for the prisoners to use the bath available at the medical unit was declined; some of the prisoners thus preferred to wash themselves at the sink in their cell rather than in the communal facilities.

Prisoners were confronted with stairs to reach various services, premises, and the facilities they needed to use. Prisoners who had difficulty with stairs frequently had to take detrimental detours when other possibilities were in theory available, including shortcuts or lifts. Anita described it, for instance, as a very tiring and distressing experience to carry heavy books on stairs.

Prison setting no. 1 was composed of several floors where the cells are on either side of long, uniform landings that seemed like stone mazes with very few orientation points. This made it very confusing, frustrating, and disorientating for some older prisoners, especially those with a high cognitive age. Several of the walking routes were found to be unsafe; in some places, the floor tiles were broken or loose, and except for the medical department, there were no support structures or handles attached to the wall. The visiting area too was situated a long distance away for most prisoners. For visitors themselves, the prison environment might also have a deterrent impact on those with high(er) functional ages. There had been an attempt to integrate art against the wall in the central hallway to provide an orientation point. Depending on what was imagined, this could potentially accentuate the 'institutional' sense of the place.

The ageing process is often accompanied by physical ailments and 'age-related symptoms' such as stiffness, joint fatigue, reduced muscle strength and endurance, and loss of hearing and vision. For most these posed few issues. For some prisoners, however, such complaints were exacerbated by the inadequate prison infrastructure, because not all places and infrastructure in the prison were easily accessible or age-friendly.

The 'lived space' of the prison then created access problems for prisoners with a high(er) biological-functional age. The large complex, spread over three levels, had an inevitable impact for two of the most common problems for functionally older prisoners: stairs and long walking distances (Couper and Fraser 2014: 12). Equally, the remoteness of the building, the withdrawal to the constrained prison cell, amounted to an amplified loss of liberty and autonomy for some older prisoners (Sykes 1958/2007: 65ff.). It should be noted that whilst an overwhelming majority were able to make their way around the prison unaided, walk reasonable distances, and manage the stairs fairly well (cf. Mann 2012), the material prison environment was undoubtedly unsuitable for prisoners functionally advanced in age. In the same vein, the lack of personal ownership and physical comfort in their material prison world arose as more generic and additional pains of imprisonment.

## Prison Setting No. 2

### Privacy: Control

The distinct architecture of prison setting no. 2 required working with small groups of prisoners. It had a spacious and accessible design whereby groups of prisoners were clustered in one-unit medical grounds and had at least two functional areas at their disposal. The prisoners were housed either in the dormitory or in one of the few private rooms. Whilst the latter had their 'own space', for the first group, the communal spaces imposed more social interaction which results in both conflict and cohesion.

## Territoriality

A minority of older prisoners had their own physical private space and the possibility of 'being alone', although Frans, one of the core participants, claimed that he has to request for his door to be locked in the evening, so as not to be disturbed by others. Most prisoners were however housed in the dormitory. These prisoners attempted to gain more control over this distinctly communal environment by marking out their territory. The personal space of prisoners was often reduced to their bed, which served as a barrier to others, as was witnessed by one of the prisoner-servants:

> I: Do you think there is enough consideration for people who prefer to withdraw?
>
> R: […] During the afternoon, he [older prisoner] also sits in front of that television, whilst he isn't actually watching, just fiddling a bit with correspondence, which is more pretext than anything else. But those are the moments that he wants to sit in the company of others and for the rest, he has very … His bed is really his cocoon. And I think he also needs that … Because he's really a bag of nerves. And if he has his cocoon, he can a little bit … And actually, it is also very stressful for him when you're saying something or asking him something over there (laughs). So, I don't do that as much. Well, handing something over to him, occasionally, that is still acceptable. (Interview Peter, March 2016)

For him and some of his peers, the primary territory was confined to the perimeter of the bed. The dormitory was very open and provided no structural material or visual barriers between spaces. There were no partitions, blinds, folding screens, or curtains, leaving these prisoners exposed to other prisoners and prison staff. This was noted by Milhaud and Moran

(2016: 180) who argue that privacy in dormitory settings is more generic than in other types of penal spaces and constitutes a further 'pain of imprisonment'. Some prisoners would pull their bedding entirely over themselves whilst lying in their beds during the day, shutting out as much environmental light, sound, and movement as possible.

Peter, one of the prisoner-servants and key figures, best summarised how the atypical environment of the gero-medical unit operated and shaped the lives of those housed here.

I think the dayroom is a very good system. I think it would be ideal, to have that in a prison, certainly if you have older people, … It also seems wise to have such a space where you can offer them activities. Then you should not always send them to another prison block to find a room, so they have that at their disposal. That is very good. It also gives the feeling of being a bit like in a real nursing home, perhaps an old-fashioned one, that might be true. But at the end of the day, those men are also a little bit old-fashioned and so that matches. If they are not disturbed, then why should we bother about it. And in the end, those guys do not care about whether there are curtains hanging or not, but if it's warm inside. They are going to complain about whether it's hot or cold and about the soup and the food and whether there is coffee, but less about at which hour and stuff. There are things that they attach more importance to, but they do appreciate a kind of conviviality ('gemütlichkeit'), which makes them forget a little bit that it's prison and I do not think that's a bad thing. […] What I do find … I think as much privacy as possible will presumably preserve human dignity and keep it intact. And I think that … if now there is a conversation taking place at someone's bed, a conversation about something medical, then the people can hear everything. I'm not saying that they… But you notice that it sometimes also leads to unmanageable behaviour of many men, which is to some extent prompted by that loss of privacy. So, how could we expect from them that they have to change underwear every day if they are in a communal sleeping area where they have to undress, whilst the whole time they embarrassed about nudity. Because that's all extremely sensitive, so I think you need something chambrette like, or even a real separation:

> […] But, I think, if they ever would convert the dormitory into ten rooms so to say, or chambrette-like situations, then you would reduce such behaviour … I think we live in a culture where that kind of erm, well, it's not that normal anymore of having that kind of contact with strangers. Even if you are sitting … You still remain completely strange for the other, because you don't have chosen to be housed together. It will benefit the interactions with each other too … I think if they all were to have single rooms, they might greet each other in the morning more, and you would also be less disturbed by each other's nighty sounds and so on. And I also think that it is good to generally have the nights and the moments that people would like to lie on or in their beds, that that would be calm moments … And to have a place that is a bit of you. I believe that people that's a fairly basic need of humans. (Peter)

By the same token, some of those in the dormitory hinted at their loss of privacy. The prison governor echoed the observation of Milhaud and Moran, (2016: 180) that prisoners had more privacy in dormitories relative to those in cells of two or three persons. However, prisoners housed in mono-cells (in both settings) responded positively to the greater privacy their penal space afforded them.

In contrast to common night space, the common day space in the living area revealed environmental efforts to establish 'intimate corners', spaces where a person can retreat. These centralised around the aquarium, the television, and tables with more comfortable upholstered seating (sofa at the television and seating within sight of the fish in the aquarium). Some of the core participants appeared to benefit from the communal day area:

> When I remain in bed, then I must be sick. There must be something wrong then. Because when I lie down on my bed—initially I tried that, especially after I've had dinner—, within fifteen minutes I must get up again because otherwise I'll start to vomit. I simply cannot lie down. (Albert)

Although prisoners denied the formal allocation of places at the (dinner) tables, many returned to the same chair to mark 'their territory' by placing their belongings on 'their table'. The following excerpt illustrates this:

> When Frans is not wandering around the hallway, he sits at his 'throne'—as he calls it, i.e. a distinct upholstered wooden chair with armrests and high backrest. (Field notes P2, January 2016)

All prisoners had a somewhat fixed spot at the table in the dayroom. This was assigned to them when they were admitted to the unit and usually stayed the same throughout their stay. The psychiatric nurse, and in his absence the prison-servant, assigned new inhabitants a spot, taking account of the prisoners' needs; they were seldom moved, as it is deemed important to them to be able to enjoy a fixed seat at the table.

There was a division between the sleeping area and the living area, and staff discouraged sleeping in the living room, but they were often seen to do so in a chair, or wheelchair, in front of the television. At times, this space was also defined through interactions, prisoners clustering around a table, for example. There were greater instances of communication and interaction in the day area, where prison staff could circulate. Prisoners shared this common space, less as individuals and more as 'a group'; they had less freedom to determine what they did, because they shared the space and facilities like the television, toilet, windows, and the refrigerator. Nevertheless, some core participants found much solace in the communal living space and spent most of their time there. Although this was dependent on being able to get in and out of bed independently, a minority experienced a heightened sense of autonomy in the common areas. For example, one prisoner even had his own lockable closet with provisions.

## Felt Space

When asked to discuss the prison environment, the prisoners referenced the variant ways in which the prison regime interacted with the infrastructure. Bobby's input supported the significance of having an open-door regime in the felt space:

> I: Have you been somewhere else before?
>
> R: Yes. I come from Prison 1 [first prison setting].

I: For how long?
R: Two months.
I: How was it over there, in relation to over here?
R: That was... a little more prison than here. Because the doors are closed.

In comparison to the facilities in prison setting no. 1, this setting was perceived by prison officers as atypical and somewhat privileged in the Belgian penal landscape.

One wheelchair-bound and partially paralysed core participant who allegedly bullied more cognitively vulnerable prisoners and deliberately soiled the toilet area with faeces was punished by being temporarily transferred to another unit without the opportunity to use the day area. This prisoner was thus segregated from the segregated unit, in what was rationalised as 'a (re)education measure' by others, as a consequence of his 'bad behaviour'. Whilst prisoners at this unit tended to experience more freedom of movement, they still lacked the feeling of being able to 'head out' and leave the prison walls from time to time.

This was elaborated upon by Albert, one of the older prisoners:

I: Is there prison staff present at night?
R: No.
I: How do you find that?
R: Erm... if it's really urgent, then there's a dead person. That person's dead. Erm... If you need care at night, and he's in the dormitory with us. Or you must have it at the door of the hallway, then they'll take care of you. But they don't come inside. So, if it's something serious wrong, that person's dead. Because, erm, to come inside at night, they must be with four or five staff members and they must call in an adjutant who must be present. So, by the time everything's in order...
R: That's erm, no system ...
I: Do you sometimes think about that?
R: Erm ... I think of it as little as possible. Hoping that it won't be necessary. Well, if we [those residing at the dormitory] pass by those individual cells at night, and we were to see someone ... who hangs there or whatever ... uh, to sound the alarm, why?! By the time that they [prison staff] are there .... (Albert)

Like the core participants in the first setting, the phenomenon of 'not feeling at home' was also reflected in the interviews:

> I: Do you try to feel at home here in any way?
> R: No, it's not my home … But I do try to make the best of it …
> I: How do you try to do that? How do you manage that?
> R: Well … My own space is my place at the table. Because I don't use the sleeping area. (Albert)

During one of the interviews in the multifunctional offices at the pavilion, the doors were noted to open inwards, which limited the circulation space for prisoners with assistive devices. When a comment on the fire safety aspect of the doors was made, Albert added that:

> Yes, but it wasn't built with fire protection in mind over here; it is built in a way that you can't leave the building.

## Autonomy

The open architectural prison design was acknowledged for the ample social and physical space for frail older and/or disabled prisoners:

> Peter the servant says that the physical aspects must be taken into account. He states, for example, that it should be borne in mind that there is enough room for wheelchairs at the table. In addition, he says, there is also the mental element (he gestures on his head); that there are also a lot of people who have psychological difficulties. He says that both things account for the need for some living space (he uses this term literally 'living space') for the people. He states that this is even more important than the physical aspect. He says that he recently discussed with a staff member about the fact that there are actually 15 inhabitants, with 13 detainees and 2 servants, and that this actually suffices. He says that if there were to be 5 more, it would be very busy (because there is a capacity of 20 people). He says that they try to take that into account in allocating seating, where they think that a person will fit in the best. (Field notes P2, January 2016)

Residents of Ward 1 who either temporarily or permanently use mobility aids were granted the extra space needed to move, take a seat, rotate, or navigate comfortably. Despite this, several older prisoners indicated that they had fallen more than once during their time at the unit. The environment evidently did not remove the risk of falling, but still enforced a degree of social control over prisoners. At night, this could still cause problems for functionally or biologically impaired prisoners who were unable to lift a fallen person, and the prison-servants were also locked up during the night.

In other respects, the atypical nature of the setting was apparent. Prison officers indicated a substantive disparity between different units in the pavilion; the gero-medical unit had a large space where inmates could move around, whilst other units were limited not only in terms of space but also that everything took place in one single space (i.e. the dormitory for most prisoners).

One of the prison officers summarised the gero-medical unit in this regard:

> There are also drawbacks, but I think that this system mainly offers advantages". "Especially for such a population" she says. She says, "you see that for example on the walk, on the other pavilions, during airing time it's crowded, there almost everyone goes walking, while here hardly anyone is going to walk, because they already have so much freedom to move". She says: "the disadvantage is that they have less privacy, for example. In a single cell you can choose what to watch on television for instance, although ... if you're housed in a cell for four prisoners (like in other units of the prison) then it's also not possible. (Field notes P2, February 2016)

## Physical Demands and Comfort

This prison had made incremental adjustments in terms of more suitable bathing and toilet facilities, hospital beds, and ramps. Wheelchair-bound prisoners had enough freedom of movement inside the unit and were able to sit at tables with the wheelchair. There were few dark and hidden spaces. It had concrete walls but apparent natural light, open spaces, and natural green materials such as peat, grass, and flowers; some prison actors described the unit as a 'poor' or 'highly secured' nursing home.

## Outside the Unit: Accessibility

Ramps were installed for wheelchair users at all entry (and exit) points, although the inner courtyard was difficult for wheelchair users as it was paved with cobbles. Only three core participants took part in communal walks in the outer courtyard, organised solely for the gero-medical unit. Wheelchair-bound and other functionally impaired prisoners needed substantial amounts of support to access outside areas. One said:

> We are with four wheelchair users and seven people in total. One of the non-wheelchair bound core-participants is also transported in a wheelchair before the walk. The psychiatric nurse pushes a wheelchair, one prisoner (50+) pushes himself with his foot and I push a wheelchair with one of the second core-participants, and another core-participant is pushed by another prisoner (50+). One of the prisoners says that they should actually have an electric wheelchair. It is quite an undertaking to walk in the inner courtyard. Prisoners need to be careful when using the steep prison ramp. The prisoner that is pushed by me helps me get him over the gutter in his wheelchair. The others are helped by the psychiatric nurse. The ground to walk on is very uneven. (Field notes P2, January 2016)

Medical services are situated at an upper floor and difficult to access. I ask one of the prisoners (50+) how it actually works when they have to go upstairs (e.g. for the dentist or ophthalmologist). This prisoner says that he can take the stairs, that he does it with his crutch. I ask what they do with people who are in a wheelchair and cannot walk anymore. He says that they [prison guards] then carry that person up the stairs, but that that is not obvious. It strikes me that this unit is actually not fully geared to these issues (when they have to go to another department or to an upper floor), because not everything is equally accessible, but they have to make do with what they've got (Field notes P2, February 2016).

Albert also commented on structural obstacles:

I: Are there any obstacles in here?

R: Yes, stairs. Stairs, and for the rest… Yes, and the gates [laughs]. Those are obstacles too, huh? [laughs]. But well, that's simply necessary.

I: I've also wondered; how does that work with your wheelchair?

R: I'm not going upstairs. I can't go upstairs, huh. But, well, I could if they [prison guards] would carry me with wheelchair and everything upstairs. But it's not allowed, because if something were to happen, the persons who carry me would be responsible… And not the prison nor the guards.

I: So, they won't do it?

R: They won't do it… Because, the risk of something happening is too high. (Albert)

The segregation unit suffered from a combination of both thoughtfulness and institutional thoughtlessness. There had been adjustments within the unit, but the unit was cut off from the services in the rest of the prison. Prisoners still faced difficulties in accessing the entire prison environment and therefore lost out on some generic activities, services, and programmes. For example, they could not reach the library nor the common visitation room.

## Individuality: Identity and Connectedness

By virtue of its nature, shared accommodation left little space for personal belongings. Only those with single cells had some personalised space:

I: Did they ever reprimand you for something?

R: No, no, no. Erm… In my sleeping area, I have nothing to … attach a picture or card. I don't have that. And so, I hung something to my closet with tape, so I wouldn't damage the closet. And then there was a prison officer, a woman, who by now has toned it down; she went and made a big fuss about that; and she came back fifteen minutes later and said 'you have to take that down, or I'm going to draw up a disciplinary report' Erm. That happened once.

I: Did you take it off?

R: I took it off.

I: So, where do you hang your pictures now?

> R: Erm… I'm not giving them a second chance anymore. That's just the way of… There have been many chefs, superiors, and officials walk past and none of them said anything. But that officer… She wanted to… How do they say that again? They wanted to become a superior and she thought she could become that. […] But I won't hang anything anymore to give them a second chance.
>
> I: Do you think you should have something to hang it?
>
> R: Yes, most have a sticky board… And at my space, there is nothing. (Albert)

One of the prison officers said the main difference with being in a nursing home was that 'they have nothing over here, not even the pants they wear is their own, so to speak'.

This historic building has poor electricity facilities, not remedied by the refurbishment. As a result, prisoners were not allowed to cook individually or have their own personal computer. Activities of this type were often provided collectively, potentially leading to conflicts of interest or needs.

## The (In)Significance for the Older Prisoner

Consistent with observations of Ball et al. (2004: 468), independence was recognised here as a physical dimension of 'autonomy' (Hofland 1990). Although the physical independence of prisoners housed in the second prison setting was hindered by their high(er) functional ages, it was compensated for by the greater freedom of mobility and a somewhat less restrictive environment. In contrast to the first setting, the physical prison environment was more oriented towards support and compensating for the declining competencies of inmates. Notwithstanding that, the general prison architecture was still poor in meeting the needs of prisoners with high(er) biological-functional dimensions of age, and the design was marked by elements of institutional thoughtlessness that went beyond the unit.

The mathematical space of the two field sites highlighted the distinctive emphasis on safety and rationalisation (Maes 2009: 65). The core participants across both settings strongly agreed that the place did not feel like home. Older prisoners in the first prison setting were housed in an age-unfriendly physical environment but could withdraw into their own cell. In the second setting, there was little space to retreat, but there was more (negotiated) control over the physical environment and a more suitable 'age'-friendly environment.

Across both settings, the comments of participants on the physical environment remained fairly limited, illustrating the privacy of the social prison environment in the lives of older prisoners. This was recognised by the prison governor and the nurse in the second setting, who denounced the antiquated nature of the prison and advocated separate rooms for inmates whilst keeping the common dayroom.

The 'material constraints' of the prison environment, defined as limitations on choice and action imposed by the physical capabilities of the human body, plus relevant features of the physical environment (Giddens 1984: 174–6 in Appelrouth and Edles 2008: 761) were prominent features of the daily life of older prisoners. Although these constraints are inherent to the prison world, they exacerbate the passivity of prisoners with higher biological-functional and psychological ages (Mann 2008: 34). Against this background, Baars warns of the danger for persons who resign themselves to the existing structures (Baars 2012: 74). A medical staff member suggested that older prisoners constitute the least empowered group of prisoners, especially those who have served long sentences and are cognitively impaired (Interview October 2014). Our findings suggest that this was not translated into structural changes nor in a greater sensitivity towards the impact of the existing structures on older prisoners.

## Notes

1. For two years, Edmund Clark photographed inside the E-Wing at HMP Kingston, Portsmouth, i.e. for eight years Britain's sole prison wing for old lifers, to explore the prisoners' experience of extended incarceration and the ordering of institutional time and space.
2. In Flanders, Belgium, the vast majority (approximately 93%) of older people still live at home on a day-to-day basis, with only around 7% of

those aged 65 and over living in a care facility (cf. De Lepeleire et al. 2013: 2). Although prisons and care facilities both entail institutional living and have in common the main characteristics of a total institution (Goffman 1961), they differ widely as to their purpose (raison d'être) and their degree of closure (Davies 1989), which is often conveyed in the physical lay-out of the institution (Goffman 1961: 4).

3. Cf. Sykes' description of the deprivation of liberty, with particular reference to the restriction of moving space
4. It must be noted that this section draws on a combination of the fieldwork and the brochure compiled by the Public Buildings Administration (in Dutch 'Regie der Gebouwen') in cooperation with the Belgian Ministry of Justice (Regie der Gebouwen n.d.: 5–61). Hereinafter referred to as 'Brochure New Penitentiairy Complex'.
5. See also Buelens (2015: 2), Gevangenismuseum vzw (n.d.) (an association founded to reveal the rich stories of vagrants formerly housed in one of the buildings).

## References

Appelrouth, S., & Edles, L. D. (2008). *Classical and Contemporary Sociological Theory: Text and Readings*. Pine Forge Press.

Baars, J. (2012). *Het nieuwe ouder worden. Paradoxen en perspectieven van leven in de tijd* (3de dr.). SWP Uitgeverij.

Ball, M. M., Perkins, M. M., Whittington, F. J., Hollingsworth, C., King, S. V., & Combs, B. L. (2004). Independence in Assisted Living. *Journal of Aging Studies, 18*, 467–483.

Beyens, K., & Maes, E. (2017). Gevangeniscapaciteit, gevangenisbevolking en gevangenispersoneel: Kwantitatieve evoluties. In K. Beyens & S. Snacken (Eds.), *Straffen: Een penologisch perspectief* (pp. 253–294). Maklu.

Bollnow, O. F. (1961). Lived-Space. *Philosophy Today, 5*(1/4), 31–39.

Bollnow, O. F. (1963). *Human Space*. Stuttgart. Retrieved from https://www.hyphenpress.co.uk/samples/26/HS_sample.pdf.

Buelens, S. (2015). *Penitentiaire inrichting Merksplas Jaarverslag 2015* [Annual Report]. Vlaamse Gemeenschap. Retrieved from https://www.departement-wvg.be/sites/default/files/media/documenten/od_jv15-merksplas_def.pdf.

Clark, E. (2008). *Still Life Killing Time*. Stockport.

Couper, S., & Fraser, A. (2014). Is the Scottish Prison Service Looking After Its Older and Frail Prisoners? In A. Fraser & M. Mellon (Eds.), *Health and (in) justice*. Retrieved from http://scottishjusticematters.com/wp-content/uploads/Pages-from-SJM_2_2_SPS-and-Older-Prisoners.pdf.
Crawley, E., & Sparks, R. (2005a). Hidden Injuries? Researching the Experiences of Older Men in English Prisons. *The Howard Journal, 44*(4), 345–356.
Crawley, E., & Sparks, R. (2005b). Older Men in Prison: Survival, Coping and Identity. In A. Liebling & S. Maruna (Eds.), *The Effects of Imprisonment* (pp. 343–365). Willan Publishing.
Davies, C. (1989). Goffman's Concept of the Total Institution: Criticisms and Revisions. *Human Studies, 12*(1–2), 77–95.
De Jong, M. (2012). *Jaarverslag hulp- en dienstverlening aan gedetineerden Penitentiair Complex Brugge—2012* [Jaarverslag]. Vlaamse Gemeenschap. Retrieved from https://www.kennisplein.be/Documents/Gedetineerden/jaarverslag_2012_hulp%20en%20dienstverlening%20aan%20gedetineerden.pdf.
De Lepeleire, J., Buntinx, F., & Schoenmakers, B. (2013). *Ouderenzorg na de zesde staatshervorming*. Academisch Centrum Huisartsengeneeskunde KU Leuven.
Gevangenismuseum vzw. (z.d.). *De strafinrichting*. De strafinrichting - Gevangenismuseum Merksplas. Accessed on 08/23/2017. Retrieved from http://www.gevangenismuseum.be/achtergrond-informatie/merksplas-kolonie/gebouwen-merksplas/de-strafinrichting
Gifford, R., Steg, L., & Reser, J. P. (2011). Environmental Psychology. In P. R. Martin, F. M. Cheung, M. C. K. MCom, M. Kyrios, J. B. Overmier, & J. M. Prieto (Eds.), *IAAP Handbook of Applied Psychology* (pp. 440–470). Blackwell Publishing Ltd..
Ginn, S. (2012). Elderly Prisoners. *BMJ, 345*, e6263. https://doi.org/10.1136/bmj.e6263.
Glover, B. M. (2012). Accommodating Older and Disabled Prisoners in England and Wales. In J. Katz, S. Peace, & S. Spurr (Eds.), *Adult Lives: A Life Course Perspective* (pp. 161–172). The Policy Press.
Goffman, E. (1961). *Asylums: Essays on the Social Situation of Mental Patients and Other Inmates*. Anchor Books.
Hoffman, M. (2014). *Foucault and Power: The Influence of Political Engagement on Theories of Power*. Bloomsbury.
Hofland, B. (1990). Why a Special Focus on Autonomy? *Generations, XIV*(Suppl), 5–8.

Horsten, T. (2013). *Landlopers: Vagebonden, verschoppelingen en weldadigheids-kolonies*. Atlas Contact, Uitgeverij.
Hovey, J. (2006). *A Thousand Words: Portraiture, Style, and Queer Modernism*. Ohio State University Press.
Howse, K. (2003). *Growing Old in Prison: A Scoping Study on Older Prisoners* (p. 49). Prison Reform Trust.
Jacobs, C., & Heracleous, L. (2001). "Seeing without Being Seen": Toward an Archaeology of Controlling Science. *International Studies of Management & Organization, 31*(3), 113–135. https://doi.org/10.2307/40397511.
Maes, E. (2009). *Van gevangenisstraf naar vrijheidsstraf: 200 jaar Belgisch gevangeniswezen*. Maklu.
Moll, A. (2013). *Losing Track of Time: Dementia and the Ageing Prison Population: Treatment Challenges and Examples of Good Practice*. Retrieved from https://www.mentalhealth.org.uk/publications/losing-track-time.
Mann, N. (2008). *Doing harder time? The experience of an ageing male prison population in England and Wales*. University of Essex.
Mann, N. (2012). *Doing Harder Time? The Experiences of an Ageing Male Prison Population in England and Wales*. Ashgate.
Mann, N. (2013). Still Life: Ageing in the Prison Environment. *ECAN Bulletin, 19*, 11–18.
Matthews, R. (1999). *Doing Time: An Introduction to the Sociology of Imprisonment*. Macmillan Press.
Milhaud, O., & Moran, D. (2016). Penal Space and Privacy in French and Russian. In D. Moran, N. Gill, & D. Conlon (Eds.), *Carceral Spaces: Mobility and Agency in Imprisonment and Migrant Detention* (pp. 167–182). Routledge.
Regie der Gebouwen. (n.d.). *Het nieuwe penitentiair complex te Brugge*. De Windroos.
Rommel, W., Declercq, A., De Clercq, J., Van Audenhove, C., & Lammertyn, F. (1998). *Tussen autonomie en geborgenheid: Dementerende ouderen en hun omgeving*. Garant.
Sardegna, J., Shelly, S., Rutzen, A. R., & Steidl, S. M. (2002). *The Encyclopedia of Blindness and Vision Impairment* (2nd ed.). Facts on File, Inc.
Sykes, G. M. (1958). *The Society of Captives*. Princeton, NJ: Princeton University Press.
Van Dijk, J. (2012, December 20). Wat is je thuis voelen? *Omgevingspsycholoog*. Retrieved from http://www.omgevingspsycholoog.nl/wat-is-je-thuis-voelen/.
Vanden Hende, M. V. (2005). *Ontgrendeld*. Academia Press.

Wahidin, A. (2000). Life Behind the Shadows: Women's Experiences of Prison in Later Life. In R. Horn & S. Warner (Eds.), *Issues in Forensic Psychology: Positive Directions for Women in Secure Environments* (pp. 53–58). British Psychological Society.

Wahidin, A. (2004). *Older Women in the Criminal Justice System: Running Out of Time*. Jessica Kingsley Publishers.

Wahidin, A. (2005). We are a Significant Minority: Elderly Women in English Prisons. *British Journal of Criminology*. Retrieved from http://www.britsoccrim.org/volume6/001.pdf.

# 4

# The Older Prisoner and the Social Prison World

This chapter invites the reader into the older prisoner's social prison world. It covers the experiences of older prisoners in terms of their interactions with various prison actors, with a focus upon the way older prisoners are perceived and treated, across an integrated and segregated prison setting. Two strands of relationships within the social world are considered here, i.e. the vertical relationships with the prison institution itself and prison staff on the one hand and the horizontal relationships with fellow prisoners on the other hand (see also Crewe 2012).

## Vertical Relationships

### Social Relations: Presence and Absence

One of the most striking observations to emerge from the data analysis was that older prisoners denounced 'the System', in ways which transcended and dictated their social and individual relationships. Similar to Carlen (1982: 123), where prisoners perceived officers as being 'pretty fair in the main' and 'just doing a job' but instead criticised 'the System'


D. Humblet, *The Older Prisoner*, Palgrave Studies in Prisons and Penology,
https://doi.org/10.1007/978-3-030-60120-1_4

they were embedded in Carlen (1982: 123). Ugelvik suggests that for the prisoner 'the System' is a euphemism or a 'catch-all term' that refers to 'the various agencies working to keep the prisoners incarcerated' (Ugelvik 2014). Many negative comments of prisoners chimed with this. Dorothy commented in the context of the recent suicide of a female fellow prisoner: 'It must not be attributed to the individuals, but to the institution' (Field notes P1, August 2014).

Whilst a minority in his study hinted at various strands of the criminal justice system (e.g. police, public prosecutor, lawyers, courts), plenty of their complaints were directed at the 'prison system', perceived as:

> pretty abstract; the faceless level that controls the prison institution, that lays down guidelines for, and affects, how the prisoners are regulated and administered every day. (Ugelvik 2014: 88)

Ugelvik points out that the 'officers, specific individuals who work on the wing, do not necessarily have anything to do with it', but at the same time, they are representatives of the system (Ugelvik 2014: 88). He focusses on prisoners' complaints to illustrate the lengthy period between concerns received and responses given. There are similarities in the attitudes expressed in his study and those described by Mathiesen (1965) in which prisoners set themselves up as 'critics of the system' who:

> accuse the System's representatives of not managing to abide by their own rules and fulfil their own goals, of being inconsistent or of complying with rules too literally. (Ugelvik 2014: 79)

Ugelvik's (2014: 78) finding that everyday life in prison was perceived as arbitrary, ineffective, and lacking in logic was echoed by the older prisoners in my sample who critiqued the illogicality that penetrated their lives. Arbitrariness was illustrated by comparing rules between prisons and even between wings. Some overtly questioned the legitimacy of the figure of the 'prison officer' as a member of a professional group yet simultaneously presented a more nuanced picture of prison officers as individuals. Many older prisoners recognised the diversity amongst prison officers. However, this was influenced by the prisoners' psychological age. Those with a high cognitive and/or low emotional age were more likely to show cognitive distortions (automatic thoughts) and tended to speak less in terms of 'the system' and more in polarised or generalised terms about prison officers and by extension all prison staff.

Ivan's description of his interactions with prison staff could be considered the epitome of the (older) inmate-officer relationship:

> I: And… how's your relationship with prison staff?
> R: Well, normal. I'm correct. If you're correct, they're correct with you. [silent pause] And relationship… In fact, I don't have a relationship with them. I'm in my cell. I'll go outside. When I go outside, then erm, I sit at the table and then we go back inside to smoke a cigarette, for instance. And then, erm, back outside, so. I don't have a problem… So 'good day', erm, 'enjoy your meal', they say. I say 'yeah, rather enjoy your ordeal'. 'And erm, no…
> I: And are they pleased with such jokes?
> R: Yeah yeah, absolutely. Absolutely. Not everyone, but the most of them are able to laugh at this, yeah. Most of them.

Or as another prisoner expressed:

> 'I leave them alone and they leave me alone'. (Field notes, P1, December 2014)

At the same time, prison officers, or prison staff in general, were also subject to 'the system'. Even if prison staff had been willing to address older prisoners' concerns, they would have been hampered by the dominant institutional culture, the origin or rationale of which was often unknown.[1] Instances of this were found inter alia: in observations where prisoners (even those with high functional ages) were required to take unnecessary detours, were disallowed from using lifts or particular bathing facilities, were denied help by staff where rules forbid certain forms of physical contact, the impossibility to obtain alternate matrasses or aids, and the very rigid night procedure (cf. previous chapter).

Older prisoners were found to have the lowest priority within the prison system. In the first setting, they fell largely outside the purview of the prison gaze.[2] Medical staff pointed out that there were 'bigger fish to fry'; prison officers emphasised that they had not received specific mandates from the hierarchy. In the second setting, the officer lowest in rank was assigned to the gero-medical unit, and staff members allocated to this unit were assigned a de facto dual role both inside and outside the unit (Field notes P2, January 2016).

## Older Prisoners and Professionals of Custody and Care

Consistent with Crawley and Sparks (2005a) who identified issues in the 'the handling of older prisoners *qua* prisoner and the recognition of need or vulnerability *qua* older person', a certain tension between prison staff and older prisoners was observed. Treatment of older prisoners was generally either indifferent or unsympathetic to the various dimensions and intersectionality of old age. The absence of any discernible difference in the relationships between prisoners and custodial staff or medical staff (especially prison nurses)[3] was striking. Similar findings were reported in England and Wales:

> many of the health care professionals are unsure as to what their role within the prison actually is, with many demonstrating a distant and non-empathic attitude which is more akin to the role of custodian than carer. (Mann 2012: 106)

A Foucauldian analysis offers a distinct take on how the hegemonic power over prisoners is deployed by the regulation of medication and the exclusion of prisoners from medical discourse by healthcare staff. Healthcare prison staff members thus 'remain the custodians of a world to which many older prisoners require access' (Sim 2002: 315; Mann 2012: 47; Mann 2008: 127; Wahidin 2004: 136ff.). Prisoners are moreover generically depicted as having a 'second-class' status and not regarded as able to 'make demands' for health care (Mann 2011: 98). As a result, some older prisoners refused to cooperate with healthcare personnel as a form of resistance, preferring withdrawal from this power relationship and accepting its adverse impact.

## The Sameness Principle and Otherness Principle

Prisons produce 'prisoners', a status which entails both objectivation and 'othering' (Ugelvik 2014: 10). The findings from both settings identified additional mechanisms that confirmed, exacerbated, or inverted the effects of this hegemonic power. Old-age prisoners were subjected to mechanisms of both sameness and otherness, albeit in varying forms.

Arguably, the social environment was found to be even more decisive than the physical environment in shaping or creating opportunities that impacted on the well-being and experiences of older prisoners. Thus, where infrastructural remedies were non-existent for an age-related process, it was possible for this to be overcome or adjusted via social environmental mechanisms. Ideas and actions centred around both sameness and otherness were central to the way in which 'older prisoners' were constructed and treated within the prison. Where the concept of 'institutional thoughtlessness' refers to the dominant discourse whereby all prisoners are homogenised by prison staff (Crawley 2005; Crawley and Sparks 2005a), we also uncovered some of the discourses and non-discursive practices whereby otherness of 'older prisoners' was established.

### Towards Sameness

#### *Institutional Thoughtlessness*

Hitherto, restrictions attached to the prison environment itself—as explained above (i.e. prisons poorly adapted to the needs of older prisoners)—have been discussed. However, institutional thoughtlessness is also rooted and sustained by the *prison praxis*. Crawley found frequent examples of staff, who, consciously or otherwise, failed to mitigate the effects of this, even when it was within their power to do so (Crawley 2005: 356). During their fieldwork Crawley and Sparks (2005a: 351) discovered a set of un-noticed injuries to older prisoners that were inflicted during everyday institutional practice, including prisoners' interactions with staff. They described older prisoners in terms of their incapacity, and invisibility, within the prison system (Crawley and Sparks 2005a: 351).

The lack of special provisions made by prison officials vis-à-vis old-age prisoners is well-documented, by a multitude of statements made by ageing prisoners similar to this one:

> They treat the elderly like any other inmate. The idea of any special treatment for the elderly is out the door. Prison is prison. How they run the prison is inhumane, the same for everyone. It means you are faced with unprofessional guards. The word elderly is out of their vocabulary. You are a prisoner, plain and simple. They treat you like any other inmate. (Cool et al. 2017)

As prisons fail to recognise the older prisoner population as a distinct group, Crawley found that older prisoners are accordingly often unable to take advantage of facilities that are in principle available, but access to which is de facto ruled out (Crawley 2005: 356, own emphasis).

The first prison setting epitomised 'institutional thoughtlessness' (hereafter 'IT'). The way the prison world was constructed subjected all prisoners to a uniform regime and the constraints of the prison design. During the fieldwork, many instances of this were revealed: the restricted use of communal rooms, the rigorous and often prohibited use of lifts and bathtubs, and the enforcement of cumbersome detours inside the prison building.

Confirming the findings of Crawley and Sparks (2005a) who suggested that prison staff strongly adhere to a sameness principle, prison officer Colleen in the first setting stated in a nonchalant way: 'Well, we treat everybody the same in here!'. Some prison officers attributed negative characteristics to older prisoners, for example, 'they're a little bit slower', which was at times refuted by the prisoners. One female prisoner (65 years old) commented 'that older people would be slower and less well performing is nonsense'. Another older prisoner stated, 'Old age rather brings about that you need more time to recuperate after physical exertion, but you can still do the same'. Prisoners' functional ages were not taken into account by prison officers.[4] This sameness principle is evidenced by the following examples:

"[…] And in the workplace, you would want a bit of lighter work, whilst those youngsters are sitting on their lazy asses and we … Because they know that you're doing a better job and don't you dare to say something because they'll say: 'Oh, but if you don't like it, there's still room upstairs'." And what's up there? "That's over here (points upwards)". Oh, right, they just imply to go back to your cell. "Yes. If the stairs are too high, they say 'stay inside'. All such things".

Similarly, Ivan, an older prisoner suffering from multimorbidity, robustly asserted the hegemonic nature of the sameness principle:

I: Are there sometimes exceptions made for…?

R: Here, the law is… erm… 'if you can't do it, you can get lost. We'll find someone else… There are enough candidates over here'. No, no, no, no exceptions are made. Because I remember the first day, I went to workplace 6, they said: 'your spot is over there'. And I said: 'do you have a chair?'. 'Gosh, no no no!' It was a prisoner, as a matter of fact, a servant, who brought me a chair without me asking him. So that was kind of him. Because you had to put towels in cellophane, which was manageable when sitting down. But when folding towels, which has to be done precisely, erm… I had too much erm… back ache.

Hitherto 'IT' has been evinced largely in terms of work and prisoners' functional age. However, the IT phenomenon penetrated various other aspects of the prison regime and the lives of older prisoners in multiple ways. Prison officer Heather was asked how she would describe 'working with' older adults, to which she replied: 'Well, it's a little bit more amiable. But for them the same laws and rules apply, they must also adhere to it'. Some older interviewees took the view that no provision was made for 'older adults' in the first prison setting:

There is absolutely nothing, nothing, nothing for older persons. They also don't pay attention to you. I ask 'why?'. Because they don't have to deal with us. It's only when they need you, for instance when there's a fundraising for charity, or something like that, because we still know how to knit, huh. Then they'll notice you.

Segregated 'gero-medical units', such as the one examined in our study, have been advocated as the preferred solution for the plight of older prisoners in generic prisons. In contrast to the prison setting no. 1, the second setting's infrastructure was regarded as more suitable for the declining competencies of its inhabitants. However, even here, the prison regime was more underpinned by IT than has often been suggested. Despite physical segregation, itself an indication of 'otherness', older prisoners were formally subject to the same prison regime and were largely perceived as a homogeneous bloc. Whilst there were elements of thoughtfulness,[5] with activities being organised specifically for this 'target group', the inhabitants were also strongly embedded into the dynamics of the general prison on the one hand yet cut-off from some services and activities offered to prisoners outside the unit on the other. Moreover, all participants in the segregated prison unit were de facto 'retired' and thus excluded from performing any form of (paid) labour. They were also unable to, and sometimes discouraged from, taking part in general activities, despite chronological and other ages similar to those in the integrated setting.

Potential tensions can arise for prison staff when balancing between 'batch processing' of inmates (Goffman 1961) and providing individual responses and solutions:

> Bernard [an inhabitant of the gero-medical unit who has been diagnosed with young-dementia] is visited by two—apparently older—persons. At a certain point during the visit, Bernard needs to go to the restroom. Terry, the prison officer who has to supervise the visitation hasn't noticed that he's been called. I say to Terry: 'I think they need you'. Terry walks over to him, and it seems that Bernard needs to go to the restroom. Terry says: 'But when you go to the toilet, the visit will end, huh'. Bernard's visitors are in the midst of drinking and eating a snack. Terry doubts but eventually says: 'Come on, it's okay'. Terry leaves together with Bernard to go to the restroom. When they re-enter, Terry says to me: 'now that's the advantage of being an older officer. Normally it's not allowed, but hey'. Terry also states that this usually means that the visit must end because they could otherwise insert or pass on something, or such. He makes it very clear that he's made an exception.

Despite a few ad hoc exceptions, it was clear that sameness was the fundamental organising principle. This became clear during a controversial moment when prison staff required a prisoner (50+) to stop having soup in a mug and eat it from a regular soup bowl instead; a decision initiated by one of the prisoner-servants and legitimised under the pretext that 'everyone is equal, and therefore must adhere to the rules'. Similarly, prisoners with high(er) functional ages (in particular the cognitive aspect) were strongly discouraged from taking part in activities outside the unit. Furthermore, the prison staff contended that the special unit had a specialised prison regime for older prisoners and activities to which they should all adhere. Whilst the data supported the validity of IT in both the integrated and the segregated prison settings, the extent to which older prisoners were affected by it was nuanced, depending on to what extent their different dimensions of age differed from the mainstream population of the prison.

### *Authority*

To exert authority[6] over prisoners remains an important argument in the professional legitimation of many prison officers in both research settings. The various age dimensions of prisoners were not found to constitute a legitimate exception to the basis of authority. Whilst this may be unsurprising in the first prison setting, it is more striking that it also applied to the second:

Albert expresses a negative attitude towards some guards. He speaks in particular about two guards who are no longer present now. He says that 'they wanted to show that they're the boss'. He says that new (young) people often have to show that they're the boss. He says that these both officers wanted to become a superior and that they did so as part of their ambition. He furthermore says that most chefs don't like to be assigned to this unit. I ask why, whereupon Albert answers: 'because there isn't enough happening over here'. Albert says that he has known a prison officer who fell asleep snoring on his desk. Albert continues: 'And when we see somebody arriving, we [bangs his fist to the table] … and then he awakes with a startle'.

Peter, one of the prisoner-servants in the second research setting, confirmed this:

> The chef [prison officer] must also show that he's tough, and exercises authority … Whilst at the same time, is that true. I don't believe that that really helps. I think there are a whole bunch of chefs who are happy to be here, for instance at the hospital or the three units that doesn't matter that much, but where that's not as necessary as at other places. And you also notice that the people who work here permanently or try to work here permanently, that they definitely are not fans of that extremely tough approach. Because they actually see that [talks quietly] … it doesn't work. Because it only escalates and that's like … Well, it's an endless battle you start with people, and you both lose.

The inmates involved in the sample demonstrated general levels of acceptance for the authority of prison officers. However, the respect for authority per se needs to be distinguished from having personal respect for prison officers. Some older prisoners constructed prison officers in less than respectful terms, describing officers as 'occupying an absurd profession' or being perceived as highly stupid. Most of the older prisoners did however recognise and accept that they were subjected to a power outside themselves.

Tysoe and Tarbuck (2014: 124) have already pointed to the relatively compliant nature of this prisoner group, which does however not imply that they cannot be regarded as 'agents' (Mann 2012; Wahidin 2004). Compliance in old age has often been attributed to illness and disorder, rendering individuals disinclined to resist existing structures (Baars 2012: 74). This questions how far this may be driven by an ageing effect (a certain level of emotional maturity) or a cohort effect (have greater respect for, or experience with, institutional authority). The older prisoners also reflected on prison officers' age and gender. Some older male prisoners perceived female prison officers more positively, equating them with more benign treatment, while others regarded this as patronising or felt that female figures of authority posed a threat to their masculinity.

Prison officers often regarded older prisoners as more compliant. Older prisoners predominantly made use of non-energetic conflict styles, such as distancing oneself from the situation, avoidance, evasion, and giving in, thus putting little or no energy into influencing or resolving conflicts with individual prison officers (see Thomas and Kilmann 2002).[7] Particularly prominent in the interview data and field notes were refraining from making comments or replying (*biting their tongue*), leaving (*retreat to their cell*), or calming the situation with a high degree of emotional fervour (*hold their wrists under cold water to cool off*). Albert illustrated this behaviour vis-à-vis prison officers:

R: […] I do not say that out loud, but I do think that …

I: Do you have to bite your tongue here?

R: That happens. Even against a chef [prison officer]. I often think … 'Ghosh, that is uncalled for'. Not all of them, but there are some that. I'm not going to say that they really are power-mad people. But there are some of whom I often think, 'now that's exaggerated'.

I: Can you provide an example of that?

R: Erm… Hmm… For instance, punishing someone who has said something. Speaking for myself, I ask myself why you have to lock up that person. Because the chef thinks that what he has said is not acceptable. Then I think by myself 'jeez, jeez, jeez'. Those are often still fairly young people. I mean the chefs, huh. With very little life experience, uh … They come into a position that they have a lot of power, a great deal of power, over people who are often much older than them. And, hm. Well, it's prison system.

I: Do you better get along with some of them? Why is that?

R: Yes, yeah, yeah. With most of them… some younger ones as well. But you can talk to most of them. With most of them, you can talk, and in a serious way. They won't act like 'I'm right' and this and that. But, well, you have to learn to live with it, huh?

I: Is that difficult? Can you accept it?

R: I can resign myself to that. Although it's often difficult. But I know that over here you better keep your tongue between your teeth [laughs]. Because you always draw the short straw. And take for instance, Frans [i.e. another core-participant]. They ask: 'Frans, are you going to take a shower?'. He says: 'No'. Erm… Then they will get their superior. You always lose out, so. […] So that you think by yourself: I will just do it and [makes a hand gesture signifying to fuck off and says 'to hell with it']. Of course, you don't actually do that [laughs], but that's what you think then.

Another older prisoner narrated a story with the following moral: 'If you cannot do or change anything about it, don't complain about it'. From the prisoners' point of view, they distinguished themselves from younger prisoners who 'attempt to challenge or defy the system'.

### *Withholding Attention/Care*[8]

In application of the sameness principle, prison staff often displayed a failure, or outright refusal, to meet obvious needs of prisoners, and—intentionally or not—overestimated prisoners' capabilities. This happened also in the gero-medical unit, as illustrated by John:

> I: Do they let you feel that you're dependent?
> R: Yes, they make you feel that.
> I: And how do they do that?
> R: [Silent pause]. Yeah, just the ordinary things... If you just ask something, they simply don't do it, you get the answer: "you can do it yourself". Although you know yourself that you can't do it. You've already tried 1000 times, but...
> I: Where does that idea come from, that they think that?
> R: I don't know...
> I: Do you also say to them that you're not able to do it?
> R: Yes. "Yes, but you should try then, you've got all the time in the world", they'll say.
> I: 'They', that is staff?
> R: Yes.
> I: Medical staff or the guards?
> R: Both.

Displaying an appropriate level of care could lead to critique by fellow prison officers, even in the special unit. One female officer recalled being reprimanded by some of her colleagues because she had 'devoted too much attention to one particular prisoner', as he was in a palliative phase and passed away shortly after, and that 'there were also numerous other prisoners'. Similarly, in the first prison setting, a female prison officer was reprimanded by colleagues who told her that she was 'far too good' and was being taken advantage of, because she had buttered slices of bread for one of the mentally ill prisoners.

Rather than acknowledging behavioural problems as medical or adaptive issues, prison staff tended to account for care-dependent behaviour in terms of power and often referred to prisoners' 'affectation'. Moreover, officers distanced themselves from prisoners viewed as 'the others', including on a physical level: in both settings prison officers refused to push wheelchairs. Officers claimed that this was formally prohibited, a position refuted by the prison governor, thus suggesting the refusal was more a matter of culture. Some prisoners internalised, and anticipated, this with comments like 'it is not their job', and the void in care became a grey zone absorbed by other actors such as fellow inmates or the psychiatric nurse.

'Claiming behaviour' from prisoners was mostly perceived as a vector of power by prison staff, as it tended to convey that prisoners are entitled to something. Prison officers in both settings commented that 'prisoners let themselves fall on the floor to get attention', which was seen as a manipulative ploy, one to be disregarded. A 'good' older prisoner in their view was one that didn't demand much physical or other attention, as this was beneficial to the proper functioning of the prison. A conversation with two prison officers elaborating on one of the core participants who had been described by them as 'an einzalgänger' (a loner):

> He doesn't pose any problems [...] and he does appreciate a small chat once in a while, and I'll do that sometimes, not every day, hey; but he enjoys it and he's not someone who will claim you afterwards.

This echoes the claims that prisoners have a 'second-class status' where they are 'not in a position to make demands' in terms of health care or other assistance (Mann 2011: 98). This particularly held true for prisoners with high biological ages, as evidenced by observations on one of the chronologically younger prisoners (50+) in the gero-medical unit. Alex complained of how he had waited a long time for a wheelchair even though he felt breathless and weak as a result of chronic lung disease:

> Alex says he was once telephoning with his son when he asked a prison officer for a piece of paper and a pen to write something down. This officer said: 'No'. Alex replied: 'What, no?!'. That's just inhuman.

In one of the heated discussions amongst prisoners in the second setting about incidents or accidents during the night, a prisoner responded to the proposal for a prison officer to periodically check during the night: 'yeah, but then they [prison officers] must also pick him up when he is lying on the floor and not just walk over him'.

Part of the rationale behind institutional thoughtlessness is the logic of treating old-age prisoners 'the same as the other prisoners'—arguably a convoluted take on equality. To some extent this explains the justification to prison officers themselves and allowed them to construct identities as meaningful professionals, or what Baars (2012: 75) identifies as 'eldercare'. Otherwise, they would be obsolete and meaningless:

> Officer Kim states that: Mind you, there are some real heavy cases [prisoners] over here at the unit. (Field notes P2, January 2016)

Which is juxtaposed with her following statements:

> Officer Kim says there isn't much happening at the unit. She says: 'that unit takes care of itself, they also eat independently and such'. She continues: 'That's actually due to the presence of two prisoner-servants; they do a lot, like distributing food'. 'But on the whole, you have to do little over here'. She says: 'It's more like a nursing home'. She states: 'Well, I'm not gonna say that there is no need for a prison officer over there, but well, you just walk in over there once in a while to see if something's going on'.

Whilst 'older prisoners' are viewed as 'different to other older adults' in what Sykes (2007) described as their lost status as 'trusted' members of society, they are also somewhat 'different to other prisoners' by virtue of old age. The following excerpt from a conversation with a prison officer on older female prisoners illustrates the intersectionality of 'older prisoners':

> You would think of some of them that they are just old ladies, but they did not end up here without reason. I treat them now just as an old lady. When I go inside now, I'll say: "Hello, Mary (in diminutive form), I'm here with your pilley [little pill]". But the day she snaps at me or gives me a blow, that's gonna be startling too.

## Towards Otherness

### *Avoidance*

The impact of high biological age became evident when becoming participants described 'avoidance behaviour' by staff, a specific ageist term (Riva and Eck 2016: 184). In both settings, some prison officers were reluctant to check in on older prisoners, which on closer investigation was prompted by fear and anxiety of finding a lifeless inmate in their cell. 'I'd rather not experience that', said one of the prison officers in the second setting. As prisoners are sent to the segregated setting on the basis of their ill-health, such avoidance behaviour by prison staff can only lead to a lack of attention and further stigmatisation.

Some prison officers equated old age with a 'non-life state', both literally and figuratively:

> Terry [prison officer] says: 'There is no life left in those guys, huh'. […] I ask Terry if he's supposed to take up his station in the day area at all times. He says that he's occasionally supposed to have a look at the dormitory. A short while later, he stands up and walks to the dormitory. He gestures to follow him. I follow him and we stand in the dormitory. He says: 'See, there's no life in those guys! There's no life left in there anymore'.

Building on the terror management theory described in the opening chapter, prisoners with high biological ages constituted a potentially potent and threatening reminder of prison actors' own ageing, which might prompt ageist attitudes and behaviour. 'Ageism' was hence identified as a pivotal element in officer-prisoner relationships.

Antonius mentioned how frequent experiences with staff members (regardless of uniform, rank, function) were met with the avoidance of eye contact and showing signs of indifference during conversations also in the second setting. Perhaps staff members felt that old-age prisoners were unable to notice their disinterest or disregarded them because they had little 'clout': a 'structured inability to protest or register complaint in an effective manner' (Crawley and Sparks 2005a: 355–358).

*Antilocution*

The term 'antilocution' here pertains to actors who:

Unsurprisingly, antilocution creates an environment where discrimination becomes normalised and acceptable (Bordens 2017). As Condé (2004: 15) has suggested, this is a ripe environment for violations of human rights.

Explicit cases of negative stereotyping of 'older prisoners' were found in both settings. The research was often received with a combination of disdain and disinterest by prison officers, which spoke volumes about attitudes to old age. This was evinced by an informal conversation between a prison officer, a prisoner of chronologically advanced age, and the researcher:

> Prison officer: 'In fact, it is not interesting to conduct research into the elderly. It would be more interesting to research young people who you can monitor and put on the right path'.
>
> Prisoner (65+): 'Maybe she likes older men' [laughs].
>
> Prisoner (65+): 'She lets herself off easy by studying older people, so she doesn't have to continue to monitor them'.

The research was received with derision, and the term 'old' was employed as a label that few staff members, as well as inmates, were willing to accept.

For most prison officers, older prisoners lacked future prospects and were 'a lost cause' in terms of both life and rehabilitation. They were perceived as less interested in the future because they had nothing left. This was a widespread and unquestioned trope. Whilst the literature has identified similar attitudes amongst institutionalised older adults as 'a signal of a problematic life situation or life experience or interruption of the functioning of the personality' (Grommen 2006: 131), later life in itself had little meaning in the view of most prison staff and was valued accordingly. Their status as 'prisoners' rendered the notion of age as dark and otherworldly. Strikingly, this detrimental view was echoed, almost verbatim, by prison healthcare staff members: 'Why should we invest in them? They are no priority for us' and 'if they don't know better at that age…'.

Negative stereotypes and images of ageing abounded. For instance, an image hanging in one of the central supervising rooms of the officers depictured the so-called Wheels of Life, a cartoon commencing with the image of a baby being pushed in a stroller by its mother, running into the presentation of various types of vehicles in between, and ending with a frail older person in a wheelchair pushed by a nurse.

Tonya, a younger female officer in P1, drew the same parallel:

> When I walk around at Department 1 [for 'long-term' prisoners], at times I have the impression that it's more of a cheap nursing home. They have food in here, they can go to the doctor, nursing, and so on. … A cheap nursing home, huh.

Her reaction was provoked by recalling an encounter with two prisoners. The first protagonist in her narrative description was described as wearing a diaper and going to the nursing station every day. She strongly stressed and repeated the word 'diaper', while expressing the utmost repugnance. The prisoners' plight was described as 'gross'. The second protagonist was also described purely in terms of his care dependence; wheelchair bound, suffering from diabetes, and having had an amputation of his toes. The most care-dependent prisoners, often those with high functional (biological and/or cognitive age), received various forms of antilocution. This may also be related to the aversion towards providing care to prisoners per se, which, as noted above, is strongly rooted in prison officer's culture.

A prison officer supervising prison labour in the first prison setting described care-dependent prisoners as unproductive members of society, while a nurse referred to them as a burden for the prison community and a high cost for health care.

In the same vein, (older) prisoners with disabilities ('the other') often became the object of ridicule and mockery (Kitwood 1997). For example, lavatorial rituals were a particular source of derision. What would be a private matter in almost any other scenario became a public matter of laughter and amusement under the prison gaze. Signs of incontinence were strongly ridiculed or disapproved of in both prison settings:

> One of the inhabitants of the gero-medical unit (50+) says that the guards don't take him into account, or that they simply make fun of him. I say: 'Are you serious?' He says: Yeah, they'll complain and make a fuss that 'you have to do this and that', and at the same time they're laughing just behind the corner: 'He did it in his pants' and so on. But those are all old people here, they can't help themselves.

The use of semi-public restrooms was also a sensitive issue in the second setting. Staff considered cleaning up vomit, urine, and faeces as an extremely demeaning work, and this task was passed on to prisoner-servants. Vertical relationships such as between prisoners and staff are largely characterised by a lack of intimacy necessary to deal properly with issues such as this (van der Geest 2005). Consequently, it was viewed as a highly problematic issue that cut across the boundaries that both prisoners and prison staff would prefer to maintain.

### *Invalidation*

Invalidation refers to the failure of acknowledging the realities of persons (Kitwood 1997). The older prisoners denounced staff strategies that invalidated their experiences, emotions and feelings. A much observed, common retort by staff members to prisoners' grievances is that 'They shouldn't have committed the crime', or 'Well, at least you're dry inside', or 'There's no one waiting for you outside so your case isn't that urgent'. Whilst some core participants were unable to express their needs at a socio-emotional level, Anita formulated it most explicitly:

> I've already mentioned that I'm lonesome. They [prison officers] don't know how to deal with it; they don't know what to respond, because they

know that that's real to a certain extent. There once was a prison officer who said, 'but well, at least you're dry inside'. Dry?! 'You're still dry', she said. Can you believe that? I said: 'Yeah, well, that's true, but at a certain age…'. Yeah, I'm not lying, that's what she said. So, I said 'myes'. Is that because they don't know how to respond to it? It's so damn hard in here because you're older. They simply don't know how to deal with you. They haven't got the slightest clue. You're just stuck with your head against the wall. (Interview Anita, September 2014)

Prison officers demonstrated a significant lack of connectedness and personal interest in those in their care, and the opinions and feelings of prisoners were largely invalidated by prison staff. As Ivan said:

R: So that's what's life is like in this hovel over here … You don't have to ask nothing, but really nothing, of the slightest intelligence from a guard. That doesn't work at all.

I: What do you mean by 'something intelligent'? A conversation?

R: [Coughs]. A conversation… at a certain level. A conversation; they then talk about their working hours and the difficulties they have with certain fellows. Which is for me … I listen to them, but it actually doesn't interest me. Hey, if you say 'yes, I don't have any contact anymore with my wife now, should I phone her?'. They are going to have their opinion about that. But most of them would say 'I don't know'. Do you understand? To say, yes, that they have some psychological understanding somewhere, that is too far-fetched. (Interview Ivan, November 2014)

This was echoed by Anita:

I could fill a book with this [i.e. bad experiences with officers]. In those four years that I've been here, I have never met a single officer who has really talked to me or has spent a little time with me; though they have done that with the younger people. There are a few who would say once: 'how's it going?'. But well… I don't even know if there have been two such officers … Well, yes, perhaps two. Yes, yes… There used to be someone over here. He's now retired, mister Marcel. And that person dropped in sometimes. It was also an older person. But, look well, he's retired now. He would say: 'Anita, what's up' and then he said: 'you look at me so strangely'. I would respond: 'Yes, sir Marcel. I've been here for four years and you're

> the first one who asks me that'. He was a little emotional at that point. I also said to him: 'Of course that I act strangely, sir Marcel'. And he was an older person. But look, he retired some months ago.

Anita felt invalidated by the prison officers' greater degree of sympathy and understanding to younger inmates. Other female prisoners felt invalidated in comparison to male prisoners. This type of issue was found to be more pronounced for prisoners with cognitive impairments, whose realities were given fewer acknowledgements of their needs and concerns, in both settings. Unlike the first setting, the prisoners in the second setting experienced greater validation of their realities from the psychiatric nurse.

*Banishment*

In both settings, older prisoners were banished from certain prison activities or programmes. In one extreme case, an older mentally ill prisoner was banished after having a cerebrovascular accident (CVA). She was admitted to the medical unit in P1 as 'she had to go elsewhere, because she took up a room for three persons'.

In less extreme cases, older prisoners were overlooked or (in)directly excluded from taking part in generic programmes, services, or activities on grounds that it took 'too much effort to include them' as officers would need additional support to participate.[9]

Whilst the segregated prison unit could be regarded as a symbolic, if not geographical, form of banishment, Peter's account illustrated how staff members banished difficult prisoners 'to the other side' in their ('hidden') discourse. In his interview, Peter elaborates on a key event, which was regarded as an instance of both banishment and antilocution:

> R: [...] This prisoner was very violent because of his anxieties. Especially towards women. And, so, a number of nurses have received a number of serious hits. And that they don't appreciate that, I certainly understand. That results in a sort of 'chargedness', that's very human. I don't have a problem with that. That's a little bit your own powerlessness and the powerlessness of someone else and the situation: how are we going to handle

> this. And you sometimes are alone, and you have to wash someone, and then there's also a chef standing there, and he won't cooperate, and the prisoner is scared and screams and throws with things. Really, that's not easy, really. But, if you then… and I won't specify myself further, but I've heard people from staff saying that… Erm, like 'it is hoped that he dies soon', with such a disdain… […]
>
> I: And such statements or remarks, are they more the exception or the rule?
>
> R: That's more the exception. I think it also fits within, and that's another aspect of it, that prison remains a very tough environment.

### *Infantilisation*

In both settings, cell doors were not perceived as real barriers to prison officers; most unlocked and opened them at their discretion and felt entitled to walk in. The same applied for sleeping and living areas in the second setting, where the personal privacy of (older) inmates was often disregarded, with officers entering the dormitory area even though it was clearly visible through windows from the hallway:

> I [the researcher] tell Peter, for example, that I always take a detour because I do not want to walk in on people in the midst of their sleeping space. Peter says: 'Yes, voilà, but prison officers sometimes have to'. Peter also indicates that that depends on the staff member. He says they [prison officers] are also humane, and that a good prison officers performs very discreetly.

A common strategy was to draw sheets or clothing over one's head. Dorothy even reversed the gaze by confronting prison officers with what has been termed a 'profaned' body of an older person (Wahidin 2004: 133) by exercising in-cell on her bed. 'Thank god I was wearing knickers', she later disclosed to the officer in question who had witnessed the scene and felt slightly uncomfortable.

Such instances have been termed 'environmental infantilisation': the public disclosure of conditions not only dehumanises older prisoners but also decreases their ability to regulate their own privacy (Salari and Rich 2001: 124).

Prisoners' medical conditions and criminal records were not officially disclosed, though most prisoners were introduced with reference to their offence or status. For example, 'he's a paedo', 'she's a drug's nanny', 'he has early signs of dementia', or 'he has hearing loss'. Prison officers also spoke out loud among each other about the prisoners' identity and their offences. In the first prison setting, it was publicly announced in the prison workshops if someone was, for example, required to see a psychiatrist. This dovetailed with the finding that prisoners were treated like children by prison staff members. This can be illustrated by examples of speech infantilisation in the second setting:

During the conversation I have with Amy [a prison officer] in the afternoon, she says: 'And we had a diabetic here, whom is no longer here now, I guess'. And he would eagerly feast himself on those small containers of jam; because it's full of sugar, huh. He liked that. And then I would say: 'No, don't do that!' 'Just like you would do with little children huh!', she says. I ask her if that would be more the task of the care staff or for them. She responds: 'yes, more for the nurses, but, well, we are closer to them, and if I see it, then I will say something about it'.

'Around 7h30 the superior officer and officer Ben come to have a look. They both stand in the opening of the hallway of the day-area. Ben, shouting, asks me, "are they [the prisoners] behaving well?"'

The 'norms of distance' existed so as not to be confronted with ageing bodies of prisoners were violated by such prisoners, and this had to be prevented by means of punitive remarks or authoritarianism (van Dongen 1997: 56):

> When Ben [prison officer] walks out the dayroom, he taps twice on the shoulder of Mickey [prisoner] and shouts very loudly in his ear: 'Pull your pants up!

Similar behaviour is also found among prison health care staff members:

The nurse drops by in the evening. She wags her finger at Frans [prisoner] and says firmly: 'Frans, watch out, eh, put your pyjama on!'. 'Yeah yeah' says Frans …

The examples of speech infantilisation in the second prison setting were more generic and oriented towards their status as a prisoner rather than an older person:

I find it striking that a prison officer says to me: 'I have also worked inside [cellular units]'. I ask: 'Inside? You mean in the sections?'. He replies: 'Yes, that was totally not my cup of tea'. 'You must have patience, huh. I didn't have that. I couldn't stand them. They are just like small children. It's always the same thing: complaining about their call credit, about their canteen they didn't receive, etcetera'.

This deprivation of adulthood is acknowledged within penology as a form of discipline for prisoners of all ages (Carlen 1982: 104ff.), whereas gerontological studies use 'infantilisation' to refer to the same thing (Salari 2005). Whilst infantilisation describes behavioural patterns where those in authority interact with, respond to, or treat an older person in a childlike manner (Marson and Powell 2014: 143), it coincides with the discipline that is exercised to bring that older person under control. It was noted that the typical infantilisation encountered in other residential settings, whereby kindness and good intent underpin such actions (Heaslip and Ryden 2013), was seldom invoked here. Rather, infantilisation was underscored by the intent to discipline the older prisoner ('to bring him to normality'). Simultaneously, older people are differentiated from children because of their life experience which has stripped from them the 'innocence' attributed to children (van Dongen 1997: 44).

The following excerpt from the field notes illustrates how certain traits were ascribed to prisoners on grounds of chronological age. Duane, one of the higher-ranked prison officers who is responsible for the pavilion in which the gero-medical unit is housed, maintained that:

'Very young and very old prisoners are the most bothersome'. He continues: 'Like toddlers who fight', and an older one would say, 'I'm done with people speaking to me like that, I have become too old for that'.

Accounts of infantilisation can be found in studies of imprisoned (Ugelvik 2014) and older (Kelly 2005) adults, although the meaning of infantilisation for older prisoners has only been hinted at (Rowe 2011). Other scholars have found that previous life experiences and patterns will buffer the negative impact of being infantilised, yet for those without a rich history of positive experiences, the negative effect was found to be synergistic (Kelly 2005: 549). When prison actors referred to prisoners' life experiences and seniority, they were simultaneously depriving them of its value by subjecting them to infantilisation—it is in the light of this that some older prisoners advocate differential treatment.

The question could also be raised whether infantilisation was more prominent and substantive in the second prison setting, where many prisoners suffered from the type of cognitive impairments that elicit views of their incompetence and childlike status (Kelly 2005: 549), alongside the obsolescence of more traditional forms of discipline. As an officer in prison setting no. 2 stated, 'Well, you can't actually punish those guys'.

### *Objectification*

The objectification of prisoners—especially those who are affected by frailty[10]—was prevalent among healthcare staff members. In both settings, prisoners were 'processed' as objects; things that were simply there to be measured, manipulated, or pushed around, without taking into account their feelings or opinions (Kitwood 1997):

> A nurse crosses me. She looks at me strangely and asks, 'what are you doing here on a Sunday?'. She goes on to say: 'hey, over here [i.e. the medical centre] there's yet an old one'. 'She had a CVA [a stroke]'. [The nurse makes gesture of paralysis with her body, hand and legs]. Referring to this prisoner's plight, she says: 'They don't know what to do with *it* (P1).

Yet another example, where a prisoner was moved with no verbal interaction:

> Troy [male nurse] leaves the prisoner in the wheelchair waiting in front of the elevator as he walks away to quickly arrange something that was requested in the meantime. Martha [psychiatric nurse] was also planning to take the elevator. I follow her. When she suddenly notices that Troy has left, she silently sneaks off behind the back of the prisoner to the door to make a door request. She acts as if the prisoner has not heard her. She then says to me, 'Well, no, I do not do that anymore, moving patients'… 'My back is broken' 'Besides, Troy is much younger than I am, and … well, he's a man'. I nod approvingly (P1).

Many instances of carceral objectification of prisoners were witnessed. For example, 'unloading' in speaking of prisoners' airing time, officers referring to prisoners as 'four pieces' (P1), or even officers comparing prisoners to animals with an inference that prisoners were of less value than horses (P2).

### *Medicalisation*

Objectification was often accompanied by medicalisation and internalised by some of the core participants. It followed that older prisoners rarely speak in terms of feelings or emotions but in terms of (somatised) complaints. Moreover, prisoners tend to be regarded as their diagnosis:

Peter says that the prison forces you to wear an armour, to adopt a mistrustful attitude. He claims that people are as well depressed over here [the specialised unit], but that they do not really express it. Peter says: 'if asked, they might say it. You might be able to get it out of them, albeit with numerous detours'. Peter furthermore says that 'we', that is, in our culture, tend to subsume our issues under some physical label. We attribute it to something physical, for instance, by saying 'I have pain over here or over there', such as back pain, rather than saying 'I'm losing heart' or 'I'm not doing okay' (P2).

This was corroborated during an interview with Antonius, one of the chronologically oldest inhabitants at the gero-medical unit. His account demonstrates the importance for prison staff members to show sensitivity when working with vulnerable prisoners:

I: What's most important in talking with someone?

R: Yes, that they listen, and their mind is nowhere else. You can tell if they are preoccupied. And I don't waste more words on it.

I: Do you notice that often?

R: Yes. Yes…

I: Do you find that bothersome?

R: Talking to you?

I: No, when people's minds start to wander. When they are preoccupied with something else?

R: Oh. Yes. But then I end the conversation myself. Then I come up with something… 'I have abdominal pain, or, I have to go to the bathroom. Or I have headache', or whatever…

I: Then you end it yourself?

R: Indeed.

I: You do not tell it?

R: No, no. But I express it. It's for them to understand.

## Professional Script

Prison staff were expected to follow professional 'scripts' related to caring and custody[11] in the 'front region' and may express dislike or anger towards the residents among colleagues in a 'back region' (Marson and Powell 2014: 148). This was hinted at by some core participants. A detachment from and the de-personalisation of some of the prisoners emerged in the analysis and was reinforced by certain staff and/or group dynamics. Peter, one of the prisoner-servants at the gero-medical unit, alluded to the notion of 'script' that contrasted with the staffs' expression of actions without an audience:

Peter says that 'staff tries to be human here'. He explains: 'I can seldom, or not often, catch them in the act of behaving bluntly; saying or doing anything blunt. But I suspect that it does happen out of here'. He clarifies: 'staff in every sense'. I surmise that he means behind the back of prisoners. He furthermore states that 'that impedes a humane treatment of a person and that it makes it difficult to be human'. Peter believes people have to be consistent in their behaviour.

In the gero-medical unit, officers—especially those who were permanently assigned to the unit—were found to be guided less by traditional forms of custody and control[12] than officers in the first setting. This was partly explained by the fact that the prison regime was more structurally embedded in the first setting than it was for the unit. Various statements of prison officers in the unit illustrate this:

> 'just let "em"' and also 'the unit actually takes care of itself'.
>
> It's actually a live-and-let-live mentality in such a unit. You can always say something or comment on 'that can of coke is wrong, that plate should not be over there', and such. But then it would become unliveable.

Also:

> The problems that arise at the unit [the gero-medical unit]; well, those are actually the easiest problems for us to solve. Those are actually small problems, for instance, what they're going to watch on television. But, that's easy, huh. Because I have the power to say: 'okay, then I'll take the television away'. And that makes them find each other and work together because they have a common goal; that is, getting the television back. And we don't have to do anything, it just happens automatically.

A similar observation was made by the prison chaplain in the second setting:

> She [the prison chaplain] says that in prison everything revolves around security. I ask her if she thinks that it's all about security in this unit (Ward 1). She says: 'That's less over here'. I ask her: 'It's less about security, but more about … what? … care?' to which she responds: 'Over here, it's more about … just letting it happen'. 'It's more a matter of just letting them go on with it' she repeats.

It is arguable that the two settings mirror each other; while the first placed greater emphasis on traditional forms of custody and order, the second was more inclined towards care and justice. Enquiring about the tensions in relation to their care of older adult prisoners elicited the following response from an officer and epitomised the attitudes towards prisoners with high(er) needs for personal care:[13]

> They are being helped, but it doesn't have to go any further than that. If it does, then they don't belong here. I ask him: 'where do they belong then?'. 'In the medical unit', he says (P1).

It was noted that prisoners with higher biological, cognitive, and/or functional ages became ever more dependent on the prison system. Unlike Aday's (2003) finding however, the older inmates in the sample generally found little institutional support from benevolent staff persons. Most old-age adults were more reliant upon fellow prisoners, or the psychiatric nurse, than on prison officers.

### *Professionals of Care*

In both prison settings, a single psychiatric nurse was specifically allocated to a distinct group of the prisoner population. The care figure in prison setting no. 1 was responsible for mentally ill offenders who are in need of care and for all other prisoners admitted to the medical centre. The 'dichotomy in her work had grown in that way'. The dichotomous nature was similarly rooted in the function of the analogous care figure in the specialised unit at prison setting no. 2 but was deployable across the entire institution. This key figure in the first setting also took on the issues of older prisoners, in an informal, unofficial, and largely ad hoc manner:

> I *ask her why they [prison officers] call her for this older prisoner's incontinence problems. She says: "Yes, it's not uncommon that they do that." First prison officers talk about it and contemplate: 'Shouldn't we call upon her sometime?' Because I go everywhere across the entire prison, and my experience, I guess… My age, my maturity.*

The prevailing culture of ageism and thoughtlessness from prison officers and medical care professionals means care tasks are often passed on to a distinct professional of care, as found in the second research setting where the key care figure had a role that was explicitly stipulated for those in the specialised unit and not older prisoners per se.

Both performed the basic tasks associated with nursing staff. In practice this was to primarily observe the emotional and medical state of patients and to report on their progress. They were also consulted over assigning prisoners a cell or place. They were trained to support a plethora of psychological and psychiatric problems and manage patients that were generally incapable of taking care of themselves. They both performed caretaking roles—insofar as they guided patients' daily living needs—although they varied in the intensity of performing these tasks. They were more akin to a therapeutic figure, organising activities for prisoners such as playing games or walking. This served as an important cornerstone in structuring the lives of those in the gero-medical unit but was less apparent in the first setting (P1).

## The Vantage of Age: 'Old Heads' and 'Quiescent Older Adults'

Some older prisoners benefitted from their advanced age in inmate-staff relationships. Those denoted as 'old heads' by Kreager et al. (2017: 16f.) were capable of building dyadic relationships with prison officers. Prison staff preferred prisoners if they were able to make their way around prison, albeit with more repulsion and distance vis-à-vis prisoners with high cognitive ages. More often these old heads held key positions in prison: sharing little snacks, decorating the office, drawing something on officers' request, and providing a listening ear.

It has already been noted that officers regarded 'good' prisoners as those that required, or sought, little attention. Albert was one of the older ones that was viewed as a proper prisoner, who was considered to 'raise the level' at the gero-medical unit:

> I: Do you think there is respect?
>
> R: Erm… Towards me…. I diff… Yes, I think so. Because there is a guard, who's no longer here. I believe he went to another prison. He once

said to me: 'you [coughs] raise the level here in the unit'. Even though I don't know wherefore.

'Older' prisoners then, who were perceived to contribute to prison life with prosocial behaviour and peer mentoring, were indicative of maturity and psychological adjustment to prison existence (Kreager et al. 2017).

Prison officers generally considered older prisoners an 'easy' population to handle; being 'quiet' and 'compliant' were traits ascribed only to old age, which according to several prison officers was reciprocated and reflected in their 'calmer' and 'at ease' disposition. By the same token, one of the prison officers argued that if older prisoners would be removed, it would have an adverse effect on the general population. For most officers, the older population posed few issues, and many officers had not given much thought to them and their needs. When their opinion was sought on the integration or segregation of older prisoners, some preferred working in a calmer section, whereas others were unwilling to work in such units.

Interestingly, Dorothy felt that some prison officers were intimidated by older prisoners and the mental resilience that she suggested accompanied the ageing process. This turned out to be especially the case when older prisoners tend to be 'unperturbed' by the prison experience and the adverse treatment by staff.

## Horizontal Relationships

### 'Fatiks': In-between Prisoner and Staff

Irrespective of the fact that many of the core participants had previously occupied (a) similar position(s),[14] the porter (also a termed prisoner-servant or Fatik in Dutch) played a key role in the social world of (old-age) prisoners. The value attached to individuals in this position was dependent on the prisoners' profile in terms of the different dimensions of age, their functionality, the prisoner-porters' own interpretation of their functionality, and the leeway that was offered to them by the setting and other actors.

Prisoner-servants were found to provide informal care to older prisoners by, inter alia, pushing wheelchairs, distributing (sometimes cutting) food, helping with clothing, cleaning cells, providing information, and acting as a preferred contact point or as a buffer. Throughout the study, prison-servants offered an ad hoc response to the types of institutional thoughtlessness noted above. However, there was no framework for this kind of informal care, and prisoners who carried out these tasks walked a fine line between the prison staff and prisoners. This was epitomised by the two prisoner-servants working and living in the gero-medical unit, who were more akin to professionals than to inmates.

Their attitude towards prisoners, whereby one prisoner-servant sees himself as an extension of the nursing staff and the other as an extension of custody, is evinced by the following excerpt:

> Peter says that Dave was struggling, especially in the beginning, because according to him [Dave] 'it's prison, and prison has to be tough, and the same for everyone'. Peter says that 'Dave has an issue with the fact that not everyone needs the same thing'. Peter argues that not everyone over there is the same and, as such, needs a different approach or treatment.

In the statement above, Peter touched upon the sameness principle that was also applied below by Dave, one of the prisoner-servants. Peter continues and elaborates further on his own view:

> I don't need that to keep me in line. I have too much respect for … Well, I just see my father and my mother, they are of similar ages as the oldest ones who are here. Erm, and, I have too much respect for people of that age. I always say that I hope that when I reach that age, I hope they won't take advantage of my weakness. I don't want that. I hope that one treats an older person with respect. That's true for most cultures that are worthy of the name 'civilisation and culture'; there has always been much respect for people who are of advanced age. And I think that there has to be enough respect. And I sometimes find it regrettable that staff—and I understand if it's only letting off steam—staff, of any category… If they slip into unrespectful behaviour. Because then you see that those humans are not seen as human anymore. And that's so important. Damn, for every single prisoner it's so important that staff at least views him as a human. Because that's

> where most prisoners have issues with, that they are not able to do that. And if you can set the right example; that's the only constructive thing that you might be able to do in a prison, to set the right example as staff. And so, they don't have many means here, they have to make do with what they've got, that's really true. But that's one thing you can do. You can, let's say, a person that has made a mistake, in any way, even if you only stole something from someone; then you've also taken away something from someone else and didn't think about the possible consequences for that other person. So, there's always something unrespectful in that.

Peter called for more control on the work he does and considered there was little attention to the way he dealt with inmates. Prison officers considered this prisoner-servant's range of extensive duties as positive, as they perceived them to be 'improper prison work' that undermined their professional status (see also Crawley and Sparks 2005b: 358). This was only acceptable as long as they did not cross the line of professional boundaries, expressed by prison officers as 'as long as he can maintain his distance' and 'stays within his remit'.

## Inmate-on-Inmate Relationships: Between Amity and Enmity?[15]

Prison requires people who have lost society's trust to live in close proximity with each other (Sykes 1958: 76f.). Penologists have tended to portray prison as an environment that is 'not conducive to building friendships' (Corley 2001: 106). The extant research on the relationships older prisoners have with other prisoners has pointed to a degree of diversity as to the value and nature of these relationships. In terms of Baumeister and Keller's typology (2011: 52ff.), older prisoners have been found to either distance themselves from other prisoners (The trying preserver: p. 54ff.), be on hostile terms with other prisoners (The misunderstood outsider: p. 64ff.), build new relationships with their fellow inmates (The conformist restarter: p. 60ff.), or be disregarded by their counterparts (The unnoticeable lackadaisical: p. 70ff.). The results in this study indicate that the relationships older prisoners make tend to be situated between enmity and amity and are marked by distance, conformity, and/or hostility.

What we found was broadly consistent with other research which found that the explicit exclusion from reference groups associated with negative connotations was even more profound for older prisoners than entering into, and/or maintaining, individual relationships (Baumeister and Keller 2011: 49). The findings thus highlight the mechanism of 'othering', whereby older prisoners housed in mixed-age groups[16] distance themselves from negatively constructed groups, such as the 'drugged', 'foreign', 'ignorant', and 'younger' prisoners who were deemed lazy and with whom they don't want to be identified. Core participants in P1 commented on (younger) prisoners stating that 'they can't count' or that 'they are too lazy to work'.

Older prisoners then attributed a certain value to life experience and were dismissive of those with less who 'think that they know it better'. This emphasis on seniority and high psychological age was at times acknowledged by other prisoners, particularly for 'old heads'. Invariably, older prisoners regarded imprisonment during later life as a highly undesirable life experience. As a result, they were disturbed by the behaviour of (some) young(er) prisoners who voiced and epitomised quite the opposite, especially when this was accompanied by pronounced machismo—the latter of which was threatening to older prisoners:

> What is more, that you're contemplating 'where did I end up?', huh? I have often experienced it during the evening, that I'm sitting outside at the polyvalent hall and that I think to myself: 'oh dear god, didn't they learn to write, didn't they learn this and that?'. Everything that comes out of their mouth, especially amongst those drugs people, is about drugs. About purchasing, and this and that. [...]. They cannot talk about anything else, because they are, let's say, from a very early age, have they become involved in it. (Dorothy)

> I have nothing against ... I'm certainly not racist, so. But we have the tendency to, at least if you're a little bit like..., if there's hair in the shower, you clean it up. Most of the hairs we found in here are black curls. But there are persons who shaved themselves in the shower. So, erm.... (Dorothy)

A common observation related to the tendency of older prisoners to spend most of their ('free') time in serenity and solitude. The observed increase in time spent alone could be attributed to what selectivity theory describes as the propensity for affective support to gain in importance with ageing. Moreover, ageing prisoners' support stems mainly from relationships with people from outside, even if there were associations with fellow detainees. The comment below illustrates the dismissive attitude of older prisoners towards the prisoner community:

> I: What's your relationship with the rest of the persons residing here?
>
> R: With the other detainees?... Well, you greet them, and that is all. So, erm, there is no further contact, or what have you.
>
> R: There are some who ask questions indirectly, mostly health related, and I don't know what else. Or just, to chat for a while, huh. And that's mostly... talk that actually means nothing to me. So, you try to talk, but after a certain while the conversation falls silent. It falls silent, because it's not very interesting. And I let it show...
>
> I: How do you let it show? Simply by remaining silent or...?
>
> R: Then erm, I don't actually listen anymore. And if one says anything, I reply: 'Come again? What did you say?'. And in most cases, they realise it. (Ivan)

Or, as stated by Larry in the first setting:

> I only have little contact with the prison guard: I say good day and that's all, not with the prisoners.

Many of the older prisoners in the sample succeeded in maintaining themselves in the social prisoner world. By virtue of their position, they were more likely to adhere to the prison code, by not interfering with or quarrelling with fellow inmates and keeping to themselves (Sykes 2007: 84ff.; Sutherland et al. 1992: 525ff.). Whilst many older prisoners sought (whether or not prompted by ageing processes) less salvation in physical violence, they tended to keep their emotions to themselves.

Placing prisoners of similar ages together did not guarantee a positive social effect. Both within and between similar age groups of prisoners, older prisoners tended to create barriers between themselves and their peers: either on the basis of offence type (particularly for sexual offenders) or on grounds of non-chronological ages (e.g. 'mentally impaired', 'weak', 'old fashioned'). The statement about another prisoner that 'he is not lucid' resulted in Ivan's reluctance to interact with him. The exclusion of sex offenders is illustrated below by Ivan about one of the prisoners he denoted as 'old' and with whom he no longer has contact with:

> I: And have you noticed a difference between the various prison units where you have resided?
>
> R: Erm… There were [coughs], for instance, no sexual offenders, paedophiles, thus. At that other section, it's packed with them for as far as I know. Here at this unit, you have one older man who erm… likes to touch children. And that's another thing I can't understand. But I have not that intent to, like some of the others… think that's overreacted; that hate. Because that's… I think that's, erm, morbid. No one would normally do that, huh?

The extant literature on older prisoners generally refers to 'age', whereas the findings here show that prisoners' social age was an important factor in contact between prisoners. In the same prison unit where the older female prisoners were held, two very divergent meanings of social old age were apparent. Dorothy stated how she found living only amongst women one of the worst things about her life in captivity. She described it thus:

> Tweet twee tweet tweet; that's what I often do here and I say to the rest, you are just like geese. And if everyone would just put their own house in order; just keep their own hand above their heads. Then it would already improve a whole lot.

Despite this, Dorothy had very good relationships with her fellow prisoners:

> I: And what about fellow prisoners? Do you sometimes find support in them?
>
> R: Pfff, I can only say one thing. Most of them say 'mommie' to me, so… They will probably obtain support from that hassle. But for the rem-

nant, erm… It's always hemming and hawing, huh. Contemplating about: 'Should I confide in her?'; 'Can I tell her that'. Erm…

I: Is that important for you to have contact with the others?

R: Just plain contact, yes. Because I'm someone who, has to, erm, feel people and see people. Yes. But to really trust someone… I have my husband for that. Yes. So erm, when there are really troubles, then… Or I write them down. Hmm… The next day I often tear it apart. Other times, I mail it. Yes yes. So… because he also knows that if I don't like something, I tell it.

R: [During the interview people produce loud noise at a distance] It's turmoil.

I: Does it happen often that there are such issues?

R: There are times that certain ones … Yes. Then they will talk to them but… I've already requested repeatedly if it isn't possible to send someone, for instance who sits at your table and with whom you have a better relationship than with others … For instance, we laugh a lot, we talk about bras and stuff… Well, the moment you received a bad phone call, for instance, why wouldn't you let someone talk to them. I had those phone calls about those deaths in the family; and then you stand there all alone. Yes. I didn't have to go to the prison governor, huh, why would I go there. But, for instance, now someone over here had troubles with her apartment and her boyfriend cheated on her while she was in here. She had to place a few phone calls. Why can't she, for instance, ask to come over to me for an hour to have a chat. Because she's able to talk to me about it. No, they won't allow that. I think that's wrong. I won't say you have to grant that to everyone. But in certain exceptional cases. Because you might be able to talk to her and in another way. But that's ruled out here. That's not allowed. So, you're not allowed to go to each other, even if it would be in the communal area here at the unit so they would be able to see what you're doing. But, well, it's impossible.

Anita by contrast stated that she is 'with no one in disagreement'. She found it dreadful to be housed among 'young people' and even described it in terms of an additional punishment:

It's erm… a heavy, heavy burden. I've got my punishment, but I have a moral punishment on top of that. Because I cannot satisfy my need to talk or just … Always sitting and listening. And if you want to say something, they won't even listen because: 'What's she saying?!'". […]

R: Older people have other topics of discussion. Yes, yes, yes, yes. And over here, they only talk about sex and undisturbed visits and about 'that person did that' and so on. It's about nothing else here.

[…]

I: Is there someone else who you can trust?

R: No one. Take Dorothy, she's older, but she's also a child. She acts like a child amongst the children. Whether she feels good about that, I don't know… I think it has more to do with belonging. I myself cannot do that. I'm 76 years, but I'm not gonna… She also does that, talk about that sex and about that… I'm not gonna do that. Of course, yes… She's part of the group then.

## Amity

Notwithstanding this, some older adult prisoners were found to benefit from one-one relationships with fellow prisoners, and older prisoners have been found to be of great benefit to other inmates. As suggested by Baumeister and Keller (2011), these individual relationships are established on the grounds of, and closely tied to, certain functions, such as sharing mutual interests, unilateral or reciprocal support, and are often dyadic relationships (Mann 2008: 68).

These results further support the idea that older prisoners seldom recognise some of their ties in prison as friendships (Baumeister and Keller 2011: 49; Aday 2003). Prisoners have been known to refer to friendly associations (Corley 2001: 106) or to acquaintances (Mann 2008: 174), rather than to true friendships:

R: That's the same as saying, making friends. […] You cannot make a female friend over here. You can be good with someone, or someone who you can get along with etc., or you can help someone. […] 'I'm hungry, I have nothing left'. Whether I do that, or… I give enough to little black people. Well, I'd rather help someone like that. You see? In that sense. But

in other cases,... Truly saying you have a connection with someone... Besides, that's something you have to avoid. Honestly.
I: And why is that?
R: Erm... To protect yourself. You build a large wall around you.

Most older prisoners involved in the study spoke of sociable contact(s) with certain individual prisoners, rather than relationships characterised by intimacy, trust, and loyalty (Aday 2003: 135). In this respect, the research has borne out Aday's hypothesis that the more an inmate views the parameters of a relationship through the prism of a prison context or routine, the less likely they are to recognise it as a true friendship (Aday 2003: 135) and contrariwise. An exception to this was found in the account of a single prisoner who spoke explicitly about his friendship with a former resident in the specialised prison unit:

R: I had become close friends with Arthur Nightingale [i.e. a former resident], but unfortunately, he's no longer here [...].
I: Why is it that you developed such good relationship with him?
R: Why did that happen? How can you explain something like that? He has always said to me, when I asked how it happened, he said, "Well, you were admitted, and I saw there was something about you, something I didn't see in someone else." But what did he see? (laughs)... that's something I don't know .... (Albert)

The juxtaposition of the prior example with the manner in which this prisoner related to with most of his fellow residents, which were positive interactions more akin to working relationships, tended to support the aberrant nature of the former:

Albert also says to me: 'My colleague is currently off to (...) [Prison no. 1], so I'm the oldest one here now'. I ask curiously: 'your colleague?'. Albert says: 'Well, yes, 'colleague'...'. [i.e. a fellow inmate].

Conversely, Dale's account on the meaning of kin and non-kin relationships in prison also mirrors Aday's (2003) differentiation between relationships that are viewed as a consequence of common participation in the prison environment on the one hand and those being purposefully selected for companionship on the other hand.

This is my family here. I consider all of them here as my family, because I live together with them. But they are not my friends. My friends are the ones I choose myself.

The results concur with Aday's (2003: 135f.) observation that people are drawn together when they meet people with similar life circumstances to their own, which held true for both prison settings. This comment by Ivan from the first research setting illustrates this when he stresses the peculiarity of meeting all kinds of people in prison:

> I: Earlier on in the conversation, you said that 'it's tough over here'. What exactly did you mean by that?
>
> R: My neighbour's a Frenchman. A notorious gangster, I must say. Erm, but very intelligent. And where he picked all of it up, I do not know. He's someone I can talk to, for instance. But I don't have a lot of affinity with the others. And they come to me to ask advice for such pills, for this and for that, and what not. And I know someone, who's actually a heavy junkie, whom I speak with. So, no problem, huh. And tough… Well, that's the tough part: that you meet very, very, strange people. I may be strange myself, but well, I don't know that about myself.

## Older Prisoners as (Grand)parental Figures

Hitherto, the inmate-on-inmate relationships of older prisoners have been considered through the lens of the 'traditional friend role'. It was observed how older prisoners could be accepted by (younger) prisoners as some sort of 'surrogate parents or grandparents'. One of the younger prisoners in P1 explicitly alluded to this notion by referring to one of the 'oldest' (both chronological and non-chronologically) prisoners, who was housed in the same unit as the narrator, as follows: 'I regard him as my own grandfather'.

In the second prison setting, both prisoners and staff spoke of a former prisoner whose name could no longer be recalled but who was dubbed *Gramps*. He was described in terms of high frailty and had passed away immediately prior to the commencement of the fieldwork. These circumstances show that older prisoners may receive support—which mostly pertains to material, social, or physical support—from other prisoners within the caregiver-caretaker nexus. This informal care partially meets the needs of (older) prisoners often left unaddressed by the prison as a result of institutional thoughtlessness. For example, small acts of generosity like pushing wheelchairs or giving other assistance for older prisoners were observed. These kinds of relationship arose on compassionate or humanitarian grounds, which was aberrant but nevertheless present among a few prisoners.

The surrogate (grand)parental relationship here opened up the opportunity for other prisoners to 'care' for persons in old age. The 'old heads' who were perceived as powerful and influential by other prisoners were afforded a certain level of protection by other prisoners. One younger prisoner in P1 commented on the status of his older cellmate: 'I reckon that others were to stick up for you if anything would happen during airtime', when discussing possible assaults on this older prisoner. The higher status that older prisoners possessed in their unit was not only attributable to the body of experience they had acquired but also to their amicable attitude and efforts to impart positive advice to younger peers (Kreager et al. 2017: 14):

> Albert says: 'I'm the oldest one here now'. I ask him: 'Do you mind being the oldest one?'. He says: 'Well, no, but it's the thought of being the oldest one, huh?'. He informs me that he has noticed that other prisoners come to him to ask about things (P2).

Advanced age, (prison) knowledge and wisdom, and time in prison or on the unit were all reasons used interchangeably by other prisoners to account for this status. These prisoners mentored (younger) inmates and capitalised on working relationships with prison staff and/or external actors and tended to display higher maturity levels when adjusting to the prison existence. Similar to Kreager et al. (2017), different meanings of age were found to structure inmate hierarchy leading to older prisoners being sought out by ('younger') prisoners. This was illustrated in Dorothy's account that most prisoners referred to her as 'mommie' and that they probably obtained support from her prosocial activities, which she described in terms of 'ado'.

A prison officer however remained unclear about the impact of older residents amongst the prisoner population in P1: 'They are called mommy-like names, so I reckon it's good for the rest of the prisoners'. With regard to the female older prisoners, one of the prison staff argued that two (of the three) female prisoners aged over 65, who were housed at the same unit in P1, 'competed' for status:

> The older females have a positive and calming effect on the rest. There are two older females housed at the same unit and that poses some issues at times. In fact, it would be better to place them each on a different unit.

Drawing on Ridgeway (1991), Kreager et al. (2017: 26) describe how the attribute of age, if correlated enough with a valued resource like advice and mentoring, will take on positive status value regardless of whether any specific older prisoner contributes to the prison unit's general well-being. This was reflected in some prison staff members' accounts of older prisoners' prosocial behaviour having a tranquilising effect on the prisoner population in P1. In the first setting, old heads had been primarily imprisoned for violent or economic crimes, whilst in the second, the old head had been convicted of a sexual offence—although the population here was more homogenous in terms of offence type.

In response to my question whether the offence plays a role, the prison officer with the highest rank at the unit says: 'Hitherto, we haven't had to remove anyone solely because of the offence'. Another prison officer says that it sometimes is also tolerated. The first officer says: 'that's actually true because they want to keep those weak ones in the population, so that they aren't the weak ones themselves'. The second prison officer confirms: 'Yes, you need the weaker ones' (P2).

The same observation also holds for the various dimensions of age. One of the permanent prison officers assigned to the second setting stated '… And that way Albert is kind of the father figure at the unit', whilst another underlined that 'the other prisoners aren't at his beck and call'. In an endeavour to expose his status, prison staff members tended to attach a high and unilateral value to the physical state of the inhabitants, which is falsified by the 'old head' of this unit:

The first prison officer ponders on the prisoner hierarchy amongst inhabitants of the gero-medical unit: 'Albert is actually somewhat the stronger one, whilst that surprises me very much because physically that's one of the weakest ones. Considering that he's wheelchair bound and not very mobile, and yet …'. This officer continues 'He's in charge of the prison unit' and reiterates 'And at the same time, he's in a wheelchair, huh, and restricted in his mobility. Isn't that strange?'.

Old heads that emerged from the group of older prisoners were the ones denoted as 'the trying preservers' by Baumeister and Keller (2011: 52–58) and are composed of both first timers (albeit with a certain amount of prison time behind them) and recidivists. In the absence of evidence to the contrary, it is reasonable to assume that old heads might be represented in 'the conformist restarters'. In contrast to older prisoners being viewed as (grand) parental figures on the basis of their vulnerability, the kind of relationship discussed here is rooted more in the vulnerability of younger or novel prisoners.

Similarly, the composition of the prison population can be considered an important factor in acquiring such status. Prisoners newly admitted, or leaving, the unit might undergo a change in dynamics. According to prison staff members, the gero-medical unit epitomised the strive to strike the right balance between prisoners that are perceived as 'weaker' and 'stronger': 'In fact, the stronger ones are a little tilted down by the weaker ones, and vice versa, the weaker ones are lifted higher by the stronger ones'. Some of the older inhabitants of the gero-medical prison unit echoed the view of Kreager et al. (2017: 14) that the ignorance and impetuousness of 'young(er)' prisoners threatened the existing order in the unit.

Old age then has the potential to increase or reduce the extent the (grand)parental roles facilitate contact with fellow—particularly—'younger' prisoners, wherein older prisoners might act more as a provider or a receiver. An overwhelming majority of the sample expressed a preference for mixed prison settings over those restricted to their own (chronological) age. Only a small minority explicitly supported the idea of setting-up segregated units. The older prisoners housed in the segregated unit were more inclined to accept this idea, although there were some who expressed no preference at all. Rather than chronological age, other types of age were found to be more significant to these inmate-to-inmate relationships, with social age shown to exacerbate, or reduce, the distance between prisoners.

## Enmity

Enmity was found in some older prisoners' accounts that related to both ethno-centrism' and 'generation-centrism'. Few prisoners tended to set themselves apart from the latter behaviour and thinking processes of the 'other generation(s)'. For women in prison, the idea of 'staying young', or at least acting young, was arguably more prevalent than for their male counterparts.

But erm, I think it's not good to have so many years in between … That generation gap. You know, there may be two or even three (!) generations of difference. So, there's no way I can be satisfied here when it comes to speak with someone or to erm … always just having to sit and listen. And if you want to say something, they won't even listen to you because they'll go like 'what is she saying?!':

> Yes, just because you're older. Just that generation gap. That's horrible! Most of them over here are girls aged twenty or in their twenties. I do like young people and they know that. That's also essential, huh. But it's for yourself that it feels so heavy. Can you imagine living at your age between people aged 70 or 80 years old for five years? Yeah, well, perhaps, okay. But it's different, huh? So, it isn't easy. And I know it's true that the young have to communicate with the old, and older has to communicate with younger and so on, but not for five years at a stretch. That's an ordeal, I tell you. So, what do you do then? You just remain silent. But what does that bring about? Loneliness. (Anita)

And she continues:

> R: But for some: you are old.
> I: Yes. Do you feel that you're not included?

R: Yes. You're not interesting to talk to. They can't handle that.
I: And why can't they handle that?
R: I don't know. Because you're older. They also see that we're wise, but many … Well, you just count for nothing.

A loss of decorum, and/or signs of disinhibition, was the dark side of ageing for most individuals. This posed issues for other prisoners and prison staff members. Due to its higher visibility, this applied particularly to the second setting where fellow prisoners were more easily disturbed by such behaviour as a result of their closer proximity. This accounted for why it was found to hold true to a lesser extent with regard to the first prison setting. As a result of physical or cognitive impairments, some older prisoners lose their (former) social esteem in ways classified as: having very little consideration for grooming or one's attire, spitting, flatulence, burping, licking out jars of jam with one's tongue, offensive language and gesturing, uncommon hoarding behaviour, and various activities involving intimate parts of the body. Such behaviour elicited the distancing of others, who found that prisoners with a loss of decorum were not adapted to their social environment. This finding is somewhat ironic; given its own lack of attention to decent personal and material conditions, the prison regime nevertheless expected high standards from its captives.

The inappropriateness of such conduct can also cause resentment within an 'older prisoner population'. This was illustrated by the following excerpt taken from an interview with Ivan, one of the older residents of the gero-medical unit, in which he commented on his 'peers':

I: Would you prefer to be surrounded by different people?
R: [Silent pause]. Yeah, somewhat more normal people. At the moment, there are a few that are not too bad. But it used to be all, erm … They were really, gosh, old people. How do you call someone like that again? … What do you become when you become really old?
I: Do you mean dementia?
R: Yes. They were really starting to get demented.
I: How did you notice that?
R: They were just drooling, sitting the whole day in their chair, and eating …

## Intimidation and Victimisation

Insofar as victimisation of older adults in prison has been addressed, it has been contained to the margins and generally rudimentary and lacking specificity. Nevertheless, the victimisation of old-age prisoners by their (younger) counterparts has often been advanced as a key argument for age-segregated prison accommodation. As is clear however, those included in the sample however expressed little overt concerns about becoming victimised.

It has been shown throughout that older prisoners were able to establish neutral or amicable relationships with their counterparts. In addition to that, the prison code (and the male code), among prisoners, ensures such conduct might remain largely undisclosed or covert, notwithstanding the fact that many were found to employ 'avoidance strategies' to avoid potential conflicts with fellow inmates (for instance, not going to the yard for walking) which reduces the risk of victimisation and unsolicited association.

Apart from being directly subjected to violence, the most common argument put forward by older adults to withdraw themselves from walking at the prison yard was echoed by Ivan:

> I: So, you never go to the yard? You've never walked on the prison yard?
>
> R: Yes, I did go to the prison yard.... In the beginning, I went a lot. I used to go to the prison yard when I was held at another unit, but... Erm, as I said, there are mostly people who, for example, don't have any tobacco. They come and beg. Another thing is about drugs. I don't do drugs. I find that very peculiar. Because, for 95 per cent, they are here because of drug-related offences. Strange, huh? The trade that you find here, and what you pay for it. Even for pills they receive here in prison... Those are free, huh. hey, that's pro deo. They still sell it.

Victimisation was primarily a concern for those convicted of sexual offences, especially those against children. It was observed that such prisoners faced an increased vulnerability to being victimised by their fellow prisoners, matching earlier studies (Viggiani 2012: 282). This extended into the prevailing stigma about 'dirty old men', where older male prisoners (mainly others-perceived age) were suspected of being child sex offenders.

Whilst this subject was not widely discussed by the core participants, a very small minority had already been physically assaulted or feared being so. In some cases, prison staff members acted to ensure that those who were more prone to victimisation were segregated from others during shower time. In the first prison setting, a single prisoner requested, and was granted, the prerogative to walk individually. Whether it can be ascribed to older adults' social status and legitimacy, their conflict strategies, the supervision of prison staff, or the underestimation of the actual number due to the sensitive nature of the topic, physical victimisation of older prisoners was the exception rather than the rule.

Data obtained from the first prison suggested that physical victimisation was linked to individual constructions of child sex offenders rather than old age. Very little data emerged about inmate-on-inmate sexual victimisation, which suggests these results ought to be interpreted with some caution. The results are consistent with those of Kerbs and Jolley (2007: 208), albeit that for this study it was found to be virtually non-existent—only one prisoner in P1 revealed that another individual housed at the same unit attempted to grope and touch him. It was striking that female prisoners did not report any physical or sexual inmate-on-inmate victimisation.

Economic victimisation was, in the view of the core participants, inherent to the prison environment. Property victimisation occurred frequently in both settings. Experience in and with prison as well as one's status in the prisoner hierarchy was a significant indicator of this occurrence. Those with less (prison) experience, and less status, were more likely to be threatened, manipulated, or fall prey to theft or extortion. Dorothy gave an account of this:

> R: Plus, the chance that they [i.e. prisoners with a problem of drug use] throw something in your cell.
>
> I: Yes. Has that already happened?
>
> R: So far, I haven't experienced it yet. I have had belongings … some of my belongings have already disappeared. That, yes. But, well.
>
> I: How is that possible? Are the doors open sometimes?

R: Yeah, it's often the case that you have to go, unexpectedly, somewhere very fast. And then you didn't close your door properly, or when the officers have opened the door and you're still at the toilet, and so on. That's the reason why I haven't placed anything of real value nearby the door. Because I think… yes, yes, one can quickly put an arm through the door opening.

In the gero-medical unit, the disappearance of material goods and possessions was often a source of conflict and disputes amongst inhabitants. Many prisoners, especially those who suffered from cognitive impairments, were unclear as to the intentionality or unintentionality of this behaviour. Inmates in the gero-medical unit lived together relatively peacefully for most of the time, although it was observed that inmate-on-inmate victimisation occurred: psychologically (e.g. insulting, making fun of, or denigrating prisoners), economically (e.g. personal goods that had been stolen, letting the air out of wheelchair tyres), and physically (e.g. make physical threats against persons, blowing, and kicking). As for sexual victimisation, prisoners and prison staff referred to 'transgressive sexual behaviour' that was considered disruptive vis-à-vis other inhabitants (such as masturbating):

At some point in the conversation, Peter talks about malicious behaviour (i.e. bullying) amongst prisoners in general. I ask him if such behaviour also occurs at this unit. Peter replies that the tires of the wheelchair of Albert regularly run flat and that he's quite certain that others place nails in those tires and so on.

Peter also says that individuals can also be aggressive at a later age. He does point out that it's important to find out what causes that aggression, such as fear.

In connection with this, we refute the idea in the literature that if 'older adult prisoners' are segregated, this automatically creates strong amical relationships between these prisoners. This is predicated upon a claim of homogeneity, namely, that the only thing they often have in common is their chronological age. However, as the findings here suggest, the older prisoner population is characterised by a heterogeneity that takes account of a more diverse understanding of 'age'—hence an attempt to homogenise (older) prisoners results in exactly the opposite.

Peter argued that the specific gero-medical prison unit is rather composed of individuals 'who occasionally clump together and then go their own way again'. To illustrate this, he reveals that collective mealtimes take up less than 20 minutes at this unit; he noticed that prisoners ate extremely quickly and left the table quickly to resume their individual activities. He suggested that better relationships would be apparent if they had their own private space:

> I ask how it is to live together over there [in the segregated unit]. Peter says that they don't actually live together. 'It is not really living together as such over here. Rather, these are individuals who occasionally interact, and not much more.' He says: 'I can compare this best to some kind of boiling liquid; from time to time it sticks to each other, but then it dissolves again, and so on'.

## Older Prisoners as 'Signifiers'[17]

For many prisoners, an encounter with old-age prisoners in prison epitomised the punitive criminal justice system. Prisoners are confronted not only with the processes of ageing, often the less attractive aspects, but also a criminal justice system in which older prisoners receive little in the way of compassion. These older prisoners then act as 'signifiers of punishment'.

The relationship with fellow prisoners can be coloured by a segregating approach towards 'older' prisoners: in the segregated setting, distinct processes of ageing were much more visible, which was more direct and confrontational for prison actors.

Talking about this issue, Peter said:

> R: And on the other hand, at the same time it is difficult to be confronted with … that you are at an older age … I mean, you just get the feeling that the person who decides that you have to stay in jail… even if you are not really mobile anymore or sick, or something else […] And as for a number of people over here, I really question how they could still be dangerous to other persons. I reckon they are at most a danger to themselves. Erm, but

well … I'm not really in a position to comment on that because I don't know everything, the whole story, so. But … I can't tell. But, uhm, then I really get scared. […] That's a very frightening thought.

I: And do you get the impression that you more directly experience this over here?

R: Yes, of course. Well, yes, the fact that you … Every day you really see people who do not only have the age … but are also less agile, have health problems … And all of that apparently is not enough to get compassion from the authorities that go about it. And they may have good reasons, but honestly, the longer I look at it, I notice that … I even feel a certain defiance towards that; which is now greater than at the beginning that I performed this work. Erm … Because I can't grasp it. Because I just think that there's way too much … well, those people cannot defend themselves either. […]

The segregated unit therefore does not safeguard prisoners from experiencing inmate-on-inmate victimisation, as is often led to believe.

## The Older Prisoner as the Other?

In examining the social and material world of older prisoners, the findings reveal that 'the System' transcends and shapes the relationships between and within the staff, the building, and older prisoners. Within this dynamic it was clear that older prisoners in particular have a low-priority status and are rarely recognised as full-fledged members of the (prison) society—a significant strand of the forms of ageism that impacts upon the lives of older prisoners. Similar to Wahidin's (2004) findings, older prisoners were subject to an ageist discourse. This was amplified by the 'sameness principle', operationalised by prison staff members and subsumed under the notion of 'institutional thoughtlessness' in the extant literature as treating all prisoners alike. Ageism in both settings then was exacerbated by the inability to recognise that treating people equally can also demand different discourses and resources.

In addition to this, the results also point to processes where ageing prisoners are constructed as 'others', which has been less visible, or underexposed, by the current literature on the ageing imprisonment nexus.

Whilst some of these have been identified as generic features of prison life, some elements strongly suggest different meanings and outcomes for older prisoners as a result of matters unaddressed under the sameness principle or as a consequence of their construction as others, which gave rise to divergent experiences. It is certainly arguable that such processes become dehumanising: eliciting a tendency to further exacerbate the more somatic and psychological injuries during imprisonment. In addition, whilst the uniqueness and individuality that is seen as a tenet of late(r) life not only remains an untapped potential in prison but also strongly undermine those with an established sense of positive well-being.

In the second setting, prison staff members operated an instrumental and ambiguous view of 'old age', which fed into a 'dual process of othering' (van Dyk 2016: 2). On the one hand, this represented prioritising the interests of the institution and its staff over prisoners, yet simultaneously on the other, having little value to prison staff, where the existence of older prisoners posed a threat in terms of their own ageing, their status as professionals—what might be regarded unconsciously as existential threats. With the exception of the psychiatric nurse, there were striking similarities between professionals of care across the settings. Due to the more stringent nature of the material and social world in the first prison setting, there was much less scope for the psychiatric nurse to give consideration to the overall older prisoner population. At the policy level, it would be beneficial to transfer elements of the communal day area and the figure of the psychiatric nurse in the second setting, to penal institutions like the first setting.

The findings identified mechanisms of both othering and sameness in older prisoners' horizontal relationships with fellow inmates. Age (and its various dimensions) has been identified here as just one of the constituting elements in the relationships with prisoner counterparts, an element that fosters either distance or attachment in inmate-on-inmate relationships. However, it was also found that fellow prisoners provided (informal) care to other prisoners, even older prisoners. Contrary to expectations, our findings exposed some victimisation of old-age prisoners in the segregated setting, and older prisoners themselves were found to internalise dominant discourses of ageism, which was reflected in the relationships with themselves or with peers.

## Notes

1. See in this regard also the contribution of Carlen (1982) wherein she pointed to the strong hierarchical organisation permeating the whole prison system, with similar adverse effects on *both* prisoners and prison staff, which is conveyed, amongst other things, in the lack of lateral channels for the transition of information.
2. Whilst most studies have looked at this from a different light, it might be argued that few prisoners—and only to a small extent—might welcome their lower visibility in prison as a means of escaping from the rigours of prison life and disciplinary power, e.g. an older prisoner who was imposed with a disciplinary sanction kept a low profile at the unit so it would pass by unnoticed that they had forgotten to implement this sanction (Field notes P1, December, 2014).
3. Whilst prisoners did mention—often poor—relationships with the prison doctor, observations on such relationships and interviews with these prison actors were virtually non-existent.
4. It must be noted that many older prisoners are still strongly willing to work, even despite some of the difficulties or limitations they are faced with, prompted by a financial or psychological rationale. Belgian prisoners are principally not entitled to receive pension benefits whilst they are incarcerated (cf. for an article on the pension benefits for prisoners under the Belgian system: Humblet and Van Limberghen 2017).
5. Thoughtfulness should be regarded as the counterpart of thoughtlessness, i.e. when the needs and sensibilities of older prisoners are taken into account.
6. In this regard, Crawley (2013: 24) regards authority rather as something 'to be acquired through the process of interaction with prisoners' by drawing on the distinction that some of the prison officers made between power and authority.
7. Their model draws on styles that are used to handle conflicts. See for more information about the drawbacks and advantages of some styles: Thomas and Kilmann (2002).
8. Many of the concepts used here were borrowed from Tom Kitwood's seminal work on dementia; in which he speaks of 'a malignant social psychology' and refers to many of the responses or strategies recognised here (Kitwood 1997).

9. See policy coordinators of the help and assistance services to prisoners at both prison settings, who are supported by the Flemish government to increase awareness and mobilise resources to invest in an older prisoner population.
10. Here regarded as vulnerability and dependency in terms of a physical, social, or mental disadvantaged position
11. Whilst the literature has identified the importance of striking a balance between custody (preventing escape), order (internal liveability), care (well-being of prisoners), and justice (legitimacy of decisions) (Morgan 1994, 1997), Tournel (2013: 277) identified custody and order as prime concerns and suggested that care and justice are instrumental to achieve it.
12. Only one of the various means to establish order. See also Tournel (2013: 91).
13. The strong tension when prison officers have to provide personal care was supported by Crawley and Sparks (2005b: 358) and one of the students who conducted research into the experiences of prison officers working with older prisoners in a Belgian (Flemish) integrated prison (Vermassen 2016). Evidence suggested it was easier for officers in the second setting to refrain from these sorts of tasks, as they were delegated.
14. It should be noted that there is variety within this function. Reference here is to prisoners who give unpaid support to prison staff in certain instrumental activities of daily prison life.
15. A relationship does not imply an affective connotation and entails a series of interactions through time between two individuals who know each other (see Tournel (2015: 81) for this elaboration on what a relationship entails).
16. Although both research settings are to a greater or lesser degree characterised by mixed-age housing, it is argued that the second prison setting maximises the concentration of 'old(er)' prisoners of chronological, cognitive, functional, and/or biological age. Whilst it is evident that both research settings differ in terms of their direct living environment and prison regime, in both settings social distancing mechanisms were witnessed against (older) prisoners.
17. Wahidin employs the term 'visible signifiers' of punishment or discipline to refer to objects (e.g. handcuffs) (Wahidin 2004: 132), whereas here prisoners themselves are the signifiers.

## References


Aday, R. H. (2003). *Aging Prisoners: Crisis in American Corrections*. Greenwood Publishing Group.

Baars, J. (2012). *Het nieuwe ouder worden. Paradoxen en perspectieven van leven in de tijd.* (3de dr.). SWP Uitgeverij.

Baumeister, B., & Keller, S. (2011). *Alt werden im Straf- und Massnahmenvollzug*. ZHAW Soziale Arbeit.

Bordens, K. S. (2017). *Just the Facts 101. Social Psychology* (2nd ed.). Inc: Content Technologies.

Carlen, P. (1982). PAPA'S DISCIPLINE: An Analysis of Disciplinary Modes in The Scottish Women's Prison. *The Sociological Review, 30*(1), 97–124. https://doi.org/10.1111/j.1467-954X.1982.tb00655.x.

Condé, H. V. (2004). *A Handbook of International Human Rights Terminology* (2nd ed.). University of Nebraska Press.

Cool, M. A., Kerness, B., Henry, J., Ross, J., Wimmershoff, K., & Frome, R. (2017). *Aging in Prison: A Human Rights Problem We Must Fix*. AFSC Prison Watch Program. Retrieved from https://www.afsc.org/sites/afsc.civicactions.net/files/documents/Aging%20in%20prison%20report%20 2017.pdf.

Corley, D. (2001). In D. F. Sabo, T. A. Kupers, & W. J. London, *Prison Masculinities* (pp. 106–108). Temple University Press.

Crawley, E. (2005). Institutional Thoughtlessness in Prisons and Its Impacts on the Day-to-Day Lives of Elderly Men. *Journal of Contemporary Justice, 21*(4), 350–363.

Crawley, E. M. (2013). *Doing Prison Work. The Public and Private Lives of Prison Officers* (2nd Edition). Routledge

Crawley, E., & Sparks, R. (2005a). Hidden Injuries? Researching the Experiences of Older Men in English Prisons. *The Howard Journal, 44*(4), 345–356.

Crawley, E., & Sparks, R. (2005b). Older Men in Prison: Survival, Coping and Identity. In A. Liebling & S. Maruna (Eds.), *The Effects of Imprisonment* (pp. 343–365). Willan Publishing.

Crewe, B. (2012). *The Prisoner Society: Power, Adaptation and Social Life in an English Prison*. Oxford University Press.

van Dongen, E. (1997). Ongelukjes en niet-ongelukjes: Infantilisering en het oude lichaam. *Medische Antropologie, 9*(1), 41–61.

van Dyk, S. (2016). The Othering of Old Age: Insights from Postcolonial Studies. *Journal of Aging Studies, 39*, 109–120. https://doi.org/10.1016/j.jaging.2016.06.005.

van der Geest, S. (2005). *Hoe vies is ontlasting? Antropologische opmerkingen.* Retrieved from https://dare.uva.nl/search?identifier=189904bb-8a89-455e-ad82-9a04e6b1f9d0.

de Viggiani, N. (2012). Trying to be Something You Are Not Masculine Performances within a Prison Setting. *Men and Masculinities, 15*(3), 271–291. https://doi.org/10.1177/1097184X12448464.

Goffman, E. (1961). *Asylums: Essays on the social situation of mental patients and other inmates.* Anchor Books.

Grommen, R. (2006). Verschuivingen in het tijdsbeleven. In A. Marcoen, R. Grommen, & N. V. Ranst (Eds.), *Als de schaduwen langer worden: Psychologische perspectieven op ouder worden en oud zijn* (1e druk, pp. 123–142). Lannoo Uitgeverij.

Heaslip, V., & Ryden, J. (2013). *Understanding Vulnerability: A Nursing and Healthcare Approach.* John Wiley & Sons.

Humblet, D., & Van Limberghen, G. (2017). Ouderdom en detentie: De ondraaglijke lichtheid van de 'bestaanszekerheid'? *Fatik. Tijdschrift voor Strafbeleid en Gevangeniswezen, 34*(154), 24–35.

Kelly, T. B. (2005). Accumulated Risk: Mutual Aid Groups for Elderly Persons with a Mental Illness. In A. Gitterman & L. Schulman (Eds.), *Mutual Aid Groups, Vulnerable and Resilient Populations, and the Life Cycle* (3rd ed., pp. 536–572). Columbia University Press.

Kerbs, J. J., & Jolley, J. M. (2007). Inmate-on-Inmate Victimization Among Older Male Prisoners. *Crime & Delinquency, 53*(2), 187–218. https://doi.org/10.1177/0011128706294119.

Kitwood, T. M. (1997). *Dementia Reconsidered: The Person Comes First.* Open University Press.

Kreager, D., Young, J. T. N., Haynie, D., Bouchard, M., Schaefer, D., & Zajac, G. (2017). Where "Old Heads" Prevail: Inmate Hierarchy in a Men's Prison Unit. *American Sociological Review, 82*, 1–34. https://doi.org/10.1177/0003122417710462.

Mann, N. (2008). *Doing Harder Time? The Experience of an Ageing Male Prison Population in England and Wales.* University of Essex.

Mann, N. (2011). Ageing Prisoners. In B. Crewe & J. Bennett (Red.), *The Prisoner* (pp. 92–102). Routledge.

Mann, N. (2012). *Doing Harder Time? The Experiences of an Ageing Male Prison Population in England and Wales*. Ashgate.

Marson, S., & Powell, R. (2014). Goffman and Infantilization of Elderly Persons: A Theory in Development. *The Journal of Sociology & Social Welfare, 41*(8), 143–158.

Mathiesen, T. (1965). *The Defences of the Weak: A Sociological Study of a Norwegian Correctional Institution*. Tavistock.

Morgan, R. E. (1994). Thoughts About Control in Prisons. *Prison Policy and Practice, 93*, 57–60.

Morgan, R. E. (1997). Are Custody, Control and Justice Compatible? The Aims of Imprisonment Revisited. In *A. Liebling, Security, Justice and Order in Prison: Developing Perspectives* (pp. 62–70). The Institute of Criminology.

Ridgeway, C. (1991). The Social Construction of Status Value: Gender and Other Nominal Characteristics. *Social Forces, 70*(2), 367–386.

Riva, P., & Eck, J. (2016). *Social Exclusion: Psychological Approaches to Understanding and Reducing Its Impact*. Springer.

Rowe, A. (2011). Narratives of Self and Identity in Women's Prisons. *Punishment & Society, 13*(5), 571–591.

Salari, S. M. (2005). Infantilization as Elder Mistreatment: Evidence from Five Adult Day Centers. *Journal of Elder Abuse & Neglect, 17*(4), 53–91.

Salari, S. M., & Rich, M. (2001). Social and Environmental Infantilization of Aged Persons: Observations in Two Adult Day Care Centers. *International Journal of Aging & Human Development, 52*(2), 115–134. https://doi.org/10.2190/1219-B2GW-Y5G1-JFEG.

Sim, J. (2002). The Future of Prison Health Care: A Critical Analysis. *Critical Social Policy, 22*(2), 300–323.

Sutherland, E. H., Cressey, D. R., Luckenbill, D. F., & Luckenbill, D. (1992). *Principles of Criminology*. General Hall.

Sykes, G. M. (1958). *The Society of Captives*. Princeton University Press.

Sykes, G. M. (2007). *The Society of Captives: A Study of a Maximum Security Prison* [1st Princeton classic ed.]. Princeton University Press.

Thomas, K. W., & Kilmann, R. H. (2002). *Thomas-Killmann Conflict Mode Instrument*. Xicom, Inc.

Tournel, H. (2013). *Van sleuteldrager tot sleutelfiguur: Een etnografisch onderzoek naar bewaarders in een gevangenis* (Doctoral Dissertation). Vrije Universiteit Brussel.

Tournel, H. (2015). *De gevangenisbewaarder: Het professioneel leven in beeld* (Vol. 7). Maklu.

Tysoe, E., & Tarbuck, P. (2014). In A. Norman & E. Walsh (Eds.), *Nursing in Criminal Justice Services* (pp. 109–130). M&K Update Ltd.
Ugelvik, T. (2014). *Power and Resistance in Prison: Doing Time, Doing Freedom*. Palgrave Macmillan.
Vermassen, T. (2016). *Gevangenispersoneel in tijden van vergrijzing – Een onderzoek naar de ervaringen van PBA's met oudere gedetineerden* (Master Dissertation). Vrije Universiteit Brussel.
Wahidin, A. (2004). *Older Women in the Criminal Justice System: Running Out of Time*. Jessica Kingsley Publishers.

# 5

# Surviving Prison in Old Age

Previous chapters have illustrated how hostile prison environments are established and maintained for those of advanced age(s). These findings revealed layers of what Crawley and Sparks (2005a) coined the 'hidden injuries of elder imprisonment', generated by the prison environment in ways that suggest that older prisoners experience reduced meaning in their lives. However, it is argued that having a purpose and meaning in life is an important resource for prisoners in coping with the daily distress of (life in) prison (Maruna et al. 2006: 176). Meaning can easily be undermined by the rigours of prison life, which has led many scholars to explore the coping strategies older prisoners employ to retain meaning by mitigating the impact of imprisonment. The following chapter bears witness to the coping strategies that are employed by older prisoners to survive in prison. It begins by unpicking the responses to stressors and disincentives in the sample, before moving on to the final chapter which explores notions of generativity in the data.



D. Humblet, *The Older Prisoner*, Palgrave Studies in Prisons and Penology,
https://doi.org/10.1007/978-3-030-60120-1_5

## The Meaning of Imprisonment for Older Adults

In considering the deeper meaning of imprisonment in late(r) life, Farber (1944) employed the notion of high self-referentiality to examine the relationship between suffering and time. He revealed that prisoners drew on the concept of 'my life' as a frame of reference that ensures younger prisoners perceive that 'life still lies ahead'; middle-aged prisoners feel that the 'best years of their lives are spent in prison'; and for 'older adults', 'life is already lived' (Farber 1944: 175). In other words, people tend to identify life with periods of their existence, the time between youth and old age, the time before prison, and the time which is to come after prison. What appears to be unbearable however is the idea that one's entire life is experienced in prison: 'one may be serving life, but one is not serving "my life"' (Cohen and Taylor 1972: 102–103).

Vanhooren et al. (2016) concluded that 'older' prisoners, denoted as such only on the basis of chronological age(s),[1] experienced greater levels of meaninglessness, no longer search for meaning in life (Vanhooren et al. 2016). Whilst it is true that older prisoners tend to do harder time, there is evidence to suggest that many, particularly those who overcome the first blows of incarceration, are able to draw upon a wealth of experiences in life, to soften the impact of imprisonment. The findings from the samples in both settings support this view, with a majority able to retain meaning in their lives despite being imprisoned. This echoes O'Donnell's observation that older prisoners possessed the kinds of skills required to soften the impact of a high pain quotient (O'Donnell 2014: 203).

This is where biological age plays an important role, as it was found that some prisoners lowered their pain quotient by constructing an expanded lifespan. One older prisoner, for example, said, 'I come from a good family tree where family members have lived long lives'; another believed to 'regain the years spent in prison' (Field notes, April 2014). Whilst other prisoners established that their time horizon had shrunk considerably to virtually nothing, thereby increasing their pain quotient, the former revealed a form of (adaptive) coping.

Similar to Vanhooren, Farber (1944) suggested that there was a curvilinear relationship between age and suffering in prison:

> with young men being relatively sanguine-there is plenty of time ahead, they will still be fit and capable when released and imprisonment is a not overly bothersome interlude and old men whose lives were 'mostly lived', are similarly unperturbed. (Farber 1944: 175)

In contrast to both Faber and Vanhooren, many of the older prisoners in this sample took a fairly optimistic view about both their present and their future. Most of them sought to find new or revive existing purposes in life, despite a sterile environment for meaning-giving. Of course, there were some who had internalised negative ageist ideas and attitudes into a lack of meaning with few positive expectations. Others, in particular those with cognitive impairments, were more oriented towards retrospective rather than prospective meaning. For the latter, prison life is imbued with meaning in retrospect, in ways that Kierkegaard described as to 'live life forwards but understand it backwards' (Moody 2002: 49).

In the second setting, Peter, a prisoner-servant in mid-to-late adulthood, reflected on the vulnerabilities and strengths of elder imprisonment:

> Young, there must… They have a whole life ahead of them, but the psychological distress [détresse], is there … And very often, an older person has, even though he's not very bright, some kind of… you might call it resignation. You might also call it wisdom, of some sorts, to deal with the fatalities of life and what not. And that wisdom is at times more clearly visible than at other times, but you see it now and then popping up. And uhm, you don't have that with younger people. That is also why they [older prisoners] don't feel the need to… they won't knock on the door when they are locked up. They won't do that. They'll say: 'Oh well, the door's locked… if the guard opens it, then we can go outside'. And a younger fellow will probably pound. That's the immense difference. (Interview Peter, March 2016)

Much was to indicate that many of the older prisoners who were involved in the research also adapted to losses, by drawing on internal structures (cf. Mann 2008, 2012). According to Stanley, a high degree of turning to intrapsychic coping was accounted for by old age: 'We are no longer in the position to rebel in the same way as younger, more vital prisoners'. This view was echoed by Antonius, who reflected on age-related differences in experiencing imprisonment:

> We have reached a certain age … Calm … We don't need all that violence anymore, it's not necessary anymore. That's behind us. But those young people … I do think that it's tough for them. (Interview Antonius, March, 2016)

Prisoners with advanced emotional age were more able to put things in perspective and/or to let things go. This dovetails with the finding that on the one hand, old age can bring about the 'resignation' (Schmid 2015; Prandi 1993: 100) of older prisoners to make old age bearable and to 'let it be' in Suurmond's (2013) terms, yet on the other, it might give impetus to 'despondency' or lethargy (Field notes, January, 2016), attitudes which undermine the search for meaning. A good reason therefore is to be vigilant against signs of defeatism amongst older prisoners.

Prison is detrimental to those prisoners who are faced with an 'existential vacuum' in later life (Frankl 1959). These prisoners are confronted with, or struggle with, meaninglessness in their current and/or previous life circumstances and no longer search for meaning in their lives. In this regard, it might be argued that the 'older' prisoners included in Vanhooren's study (Vanhooren et al. 2016: 7ff.) might be at the top of the Farber's curve, where nothing is expected from the future, and simultaneously believe that their best years have already passed. In addition, it remains unclear as to how spending (the bulk of) life outside of prison might also affect the search for, and experience of, meaning in elder imprisonment. In this regard, one older prisoner stated that: 'We've already had our lives' (Field notes P1, April, 2014). Antonius commented 'I always say that it [i.e. being imprisoned] is a big difference for young people. Their life is…, is there, you know?' (Interview Antonius, March, 2016).

Whilst the act of generating little or no expectations in later life might be comforting or 'cathartic', acting as a buffer against disappointment, this might also suggest internalised forms of ageism, which in turn compound the loss of resources for older adults by ignoring opportunities to rebuild, or preserve, resources, denoted by Segerstrom (2006) as coping of the 'ineffective kind'. Thus, most old-age prisoners do not show defiance vis-à-vis the prison experience in the same way as they sometimes do towards the ageing process.

Older prisoners, then, gain or lose meaning, or the search for meaning. It begs the question, however, as to which impoverishments and enrichments emerged within this old-age prison nexus to keep older prisoners' lives meaningful within the conditions of the prison environment.

## Meaningfulness and Coping[2]

Coping implies an adverse, stressful, or difficult circumstance that is managed. Whilst it is strongly arguable that imprisonment comprises a highly stressful experience and situation for most prisoners (Vandebosch 2004: 205), the findings suggest that its appraisal, and the degree to which the stressor is experienced as stressful, is also influenced by various 'age dimensions'. Thus, the following appeared to have an impact: meaningful relationships with the self and others; a body and mind that is liable to change; a confrontation with finitude, all external or internal events; or signals that draw attention to, or are reminiscent of, the existential process of growing older (Baars 2012), possibly altering the traditional 'pains of imprisonment' and the responses as a result thereof.[3] This is where the notion of 'coping' amongst older prisoners was largely comprised of emotion-focused coping, where the emphasis lay in 'changing the meaning of the stressor and/or transferring attention away from it' (Pierce and Shanan 2017: 30; Lyon 2012: 9–10).

## Acute and Chronic Stressors

The distinction between chronic and acute stressors that prisoners are exposed to was significant here (Massoglia and Pridemore 2015; Vandebosch 2004: 206). When asking core participants about 'situational triggers' in prison, such as receiving bad news, they often indicated how the impact of certain events was flattening, whereby prisoners of fourth age seem most well positioned:

> I: How do you cope with bad news?
> R: Ehm. I haven't received any bad news yet. Al the persons who could have passed away, I've already lost before I came here, so.... (Interview Albert, February 2016)

Conversely, older prisoners might be subjected to much chronic stress if they are unable to construct an integrated life review because of their imprisonment. Similarly distressing life events (becoming ill; losing a significant other, such as wife or parent, and not being able to say goodbye to them; being absent at the funeral due to a lack of (timely) permission, or in, what they perceive as, dire circumstances, such as being *handcuffed*) cause prisoners much acute distress and chronic hidden injuries.

The following draws upon Mann (2012: 66ff.), to scrutinise ageing prisoners' external and internal structures. The discussion centres on primary relationships with (non)kin on the outside, and their prison activities and pastimes on the inside, and is followed by an examination of their intrapsychic ways of coping.

## External Coping

In structuring the responses of old-age adults to the prison experience, well-established definitions of external coping have been drawn on and refer to coping resources external to the person, such as things that people can access and use to deal with the stressor (Barkway 2013: 240). An analogous structure was also recognised in the work of Mann (2012: 66ff.), who creates a dichotomy between older prisoners' external and internal coping structures. The former refers to all the structures that exist and which operate more or less autonomously of the prisoner, whilst the latter is comprised of the prison structure and structure of outside society (Mann 2012: 69f.).

## Outside World Relationships[4]

For most of the older prisoners in the study, their outside support systems had dwindled over time, leaving mainly single member(s) of kin giving a significant focus on close family ties, distant family members, and a few non-kin relationships. In most cases, there was still some form of contact but little face-to-face contact with these members. Contact was primarily indirect, often with sole significant person(s), and through telephone and written correspondence. Few were able to maintain close or face-to-face contact with their spouse, and even fewer received conjugal visits in prison. However, the emergence of smaller social networks for older prisoners is not in itself considered alarming. Socioemotional selectivity theory suggests that older adults, with a perceived limited time horizon, narrow their base of social interaction (Carstensen 1992). As the intensity of broader contacts decreases, and the sources of support from outside shrink, older individuals can become more vulnerable to threats related to their sense of meaning.

For the most unfortunate prisoners, their social network will almost completely disintegrate, especially so for those with cognitive problems. This may be attributed to the nature of the offence (sexual or intrafamilial offending) and/or the interaction of chronological and social age with the effects of incarceration.

In one of the interviews, Peter (50+) touched upon several significant findings throwing light on losses and gains. First, he corroborates that relationships with significant others are of great value to older prisoners in surviving the prison experience. Secondly, if a close relationship with significant others remains after imprisonment, there is a high likelihood that such relationships are strong enough to survive, despite the prison experience.

> I: Is it harder doing time later in life?
>
> R: I reckon the problems are different. And how it would be when I'm bound to stay here yet much longer, you wouldn't know… I don't think there's such a thing as 'a good age' [laughs]. I think that that… I myself am convinced that… Well, I understand that there exists something like detention, but it must be as short as possible to be meaningful and it

> must result as quickly as possible into a sort of mandatory care. I do believe that. And that that applies to all ages, but I think the problems are completely different. What you see amongst young people is that they are devastated by the relationships that fall apart, which is not the case with someone of 70 years of age; if there are still relationships, they will be stable enough to survive that. That's also simply what you see happening With the few people who still have relationships, it remains stable. Because that is already so scarred by life that it mustn't be destroyed by an imprisonment. But the problems are different, I think that … I can imagine that there is actually a lot of really hidden and lonely sadness, that we never get to see. And I'm actually afraid of that. Because your support figures will be lost for a large part.
>
> R: For now, I still have my parents, my brothers and some friends. But if I sit here for another ten years, my parents will certainly not be there anymore. Um … And … I mean, your world shrinks, and shrinks, which you cannot compensate even though you need it. And what actually keeps you alive, inside, is what's still left outside. And that is why a long prison sentence is so detrimental, because it devastates everything that is still outside. It absolutely destroys everything. It's like weed killer. It's so all-encompassing… It's not only the weed that's gone, but everything is gone, huh? Life is broken. And that's not good about it. (Interview Peter, March, 2014)

Consistent with Mann (2012), the importance of the family structure from which to draw hope and strength was a constant source of discussion amongst the participants. It was also mentioned that it was not easy, nor always bearable, to maintain relationships with family members (e.g. emergence of artificial feelings, not wanting to ruin the rare moments with family, the 'putting on a front' which is typical amongst prisoners) (Laws 2016; Aday 2003). Exposure to outside relationships may also cause additional pain or distress as they serve as a painful reminder of what is 'not fully reachable' or 'what isn't there anymore' (Mann 2012). Moreover, the proximity of social network members was found to be particularly troublesome due to the remoteness of the prisons and the high(er) ages of family members themselves (e.g. often parents of older prisoners have already passed away).

Tony, an older prisoner who was admitted later in life and served a shorter prison sentence (<5 yrs) encountered strong feelings of embarrassment and shame, prompting him to conceal his detention as much as possible:

> R: [I have very little contact with others].
>
> R: I'm ashamed…
>
> I: Why are you embarrassed? Because you are here…?
>
> R: Because I… Well, yeah. My family doesn't know it. My children don't know it. Only my [friends] know it but they don't come visit me, huh. Ashamed of myself. That I did something stupid. A stupidity. Now I have to pay.
>
> I: Are you ashamed because they might talk about you? Or why?
>
> R: I've said it. My honour is defiled. My heart is hurt. I also still have a little bit of pride, huh.
>
> I: Yes.
>
> R: This is logical, hm?
>
> I: Yes. I only asked because I wanted to know what's important to you. And what used to be important to you.
>
> R: Your family, huh. Yes. You live to build a family. At the end of the day that's the aim of everyone, isn't it? (Interview Tony, October 2014)

The majority of prisoners in the first setting had at least one person (often a close relative) with whom they were still in touch with. For most of the prisoners in the second setting, their social networks had thinned to being virtually non-existent. A higher concentration of prisoners with more vulnerable profiles (in terms of higher bio-functional and cognitive ages and criminal profile) might account for the disparity amongst settings. Maintaining proper relationships with children, or grandchildren, was even rarer. When it did occur, it happened with the consent of parents of grandchildren and commonly took place through letters and telephone calls.

Prisoners without social networks are at greater risk of isolation, loneliness, and a loss of meaning in life. The findings here corroborate previous findings that smaller, very close social networks increase in emotional value to the older inmate (Bond et al. 2005). However, although older prisoners generally feel emotionally close to relatives and can rely on existing dyadic relationships for social support (Krabill and Aday 2007), relationships with outside kin are not only a source of support but also a source of pain. Contact was highly regulated and restricted by the prison environment: modes of contact, communication, and correspondence, frequently hampered by the limitations of functional age and not remedied by the institution.

The most detrimental mechanisms were partially alleviated by giving older prisoners the opportunity to draw on new structures from outside society, epitomised by outside volunteers. In both settings, volunteers played an important role. In the segregation unit, interaction with a recurring group of volunteers took place in group(s) and was more centralised around common activities, whereas in the integrated setting, it was largely sporadic individual visits, or correspondence, with volunteers.

That is why it is good that those volunteers come around over here. You see that, that the social contact of people who on a sufficiently long basis… what that mainly compensates, I think, is that the family ties that are not present anymore amongst a lot of people, and those volunteers are an ersatz for those lacking ties, but a good ersatz. And for a number of years now, I have already experienced it myself, at Christmas, the people of the Egidius community, who come to organise a meal and take care of that, that is practically the ersatz of a family celebration. They also come with a group of which some return each year, so there are some people they still know, and that gives a little bit comfort in that very difficult end of year period, because it's sure that… They all let it show a lot less, but I am convinced that it means a great deal to them. For instance, it shows because normally a meal takes less than twenty minutes. Hey, and that's quick! They eat more quickly than the littlest of children [laughs]. I don't understand that they can eat so fast, but well. […] (Interview Peter, March, 2016)

Anita is one of the few prisoners who explicitly and extensively elaborates on loneliness in prison (because of her exclusion from the general prisoner population). She has an emotional need to converse with others and has few remaining relationships with people from outside, which she compensates for with the help of volunteers (which is grounded in both parties' religious faith) (Interview Anita, September, 2014).

> But erm, it's no laughing matter, huh. No no no… And over here, visitation… I don't receive any family visits. […] There's a lady that comes. A volunteer. I'm very happy with that, but you know… You talk differently to family. (Interview Peter, March, 2016)

This new outside structure formed an indispensable link with outside, whilst it did not link the older prisoner with his pre-prison life as former external structures do.

## Prison (Pas)Time: Prison Time and an Ageing Body

Prison studies have underlined the importance of participating in prison life as a source of finding meaning, either by self-improvement or helping others (Willis and Zaitzow 2015: 562). The significance of new roles in the penitentiary institution is briefly touched upon by Leigey and Ryder, who established that the feelings of prisoners serving life without the possibility of parole ('LWOPs') that their lives were meaningless were countered by participating in educational and vocational programmes which provided opportunities to learn and develop new skills. Several men earned a high school diploma whilst incarcerated, and a few had taken college courses. Moreover, the studied men became actively involved in activities designed to help others (e.g. prison ministry, AIDS peer education, and legal advocacy) (Leigey and Ryder 2015: 737). However, whilst the prison setting did not deliberately exclude older prisoners from taking part in such activities, it has been noted previously that those with higher functional ages were often de facto ruled out of such formal programmes, due to the social or material prison environment.

Research into meaning-giving shows that physical integrity is increasingly important with ageing (Kuin and Westerhof 2007). For those with a high(er) bio-functional and/or cognitive age, age-associated decline in everyday competence due to chronic illness interferes with the maintenance of activities that have a purpose in life. Moreover, recent international research highlights that older prisoners show specific physical and mental health needs (Carlisle 2006; Yorston and Taylor 2006), including poor mobility, impaired vision and hearing, and depression (Crawley 2005). More tentatively, Baumeister and Keller (2011: 45) stated that the ageing inmates (60+) who were included in their research were all faced with (mostly) physical impairments. In our study, John even indicated that his severe physical impairment (paralysis as a result of a stroke) before being admitted to prison had constituted a greater stress than being imprisoned (Interview John, March, 2016).

Building on the work of Wahidin, who conceptualised the embodiment of the experience of imprisonment as prison time 'inscribed' on the body (Wahidin 2002: 182), Moran pointed out that the physical body continues to age, and this ageing process is observed and experienced corporeally (Moran 2012: 571). This picture is echoed by an older prisoner (80+) stating that: 'old age is not only a prison, but in itself a punishment' (Humblet and Decorte 2013). Strongly related to older prisoners' second meaning of biological age, Wahidin (2004: 182) states that time, or 'age', (Wahidin 2004: 73), is used against older adults in prison as an additional form of 'punishment' and is in turn re-appropriated by them through techniques of resistance. Prison time is inscribed in the body and thus becomes 'body time', with the ageing body serving as a constant reminder of limited time (Wahidin and Tate 2005; Wahidin 2002: 182). Some of our findings though reflect the sense that older prisoners resist inscriptions (of prison time) on their body, by taking care of their physical bodies, masking their impediments, keeping fit, walking/designing their own forms of exercise/fitness/exercise in their cell, and buying extra ('healthy') food, even when structural strategies, policies, or programmes are lacking. Most prisoners strove to mitigate the effects of imprisonment as much as possible, even when confronted with many obstacles. 'I always say: "I did not come into prison sick and I do not want to leave prison sick"' (Interview Dorothy, October 2014).

## Activities and Time: Speeding Time's Passage in Prison

O'Donnell (2014: 181) has indicated in his seminal work that:

> Time repays close attention by slowing down and speeds up if it is ignored, which becomes intensified with advancing years. The reason for this is that for a ten year old, one year is one-tenth of his or her life, whereas for a fifty year old the same interval is only one-fiftieth of life. Also, an individual has most new experiences early in life and over time many of these become routinised and are no longer worthy of notice; the days meld, the weeks merge. As we get older and there is less variety in our environments, time seems to speed up. This process is accentuated in prison where a backward glance reveals that time has flown and there is little to show for it. However, during a highly stressful or highly eventful situation the lived experience of time slackens.

Overall, many older prisoners in our study felt that time passed somewhat quickly in prison. A great deal of the core participants appeared to succeed in ignoring time whilst keeping busy, seeking refuge in activities (albeit ones that took place within a restricted world and the privacy of their cells). Older prisoners were engaged in, inter alia, fitness, writing, crafting, reading, solving crossword puzzles, drawing, listening to classical or music from their time, watching television, and for approximately half of the sample, this also included prison work—identified as important for older prisoners.

> I: How do you experience time here?
>
> R: It goes well ahead. That's true, but well. You wake up very early, you go to work, when you return it's lunch. I go back to work, not all people do that, but I go twice. In the afternoon as well. So… Then at the evening, yes. There's always something to do, huh? So, that's… Time goes well ahead. Often that you think 'gosh'… Plus those visits. (Interview Dorothy, October, 2014)

In their work on meaning and purpose in later life, Bass and Weiss drew our attention to 'drift'. There can easily be drift: letting things happen, deciding on activities one at a time, and so establishing a routine that is good enough, but not at all chosen (Weiss and Bass 2002: 5).

In the specialised unit, the strive for a life of drift is structurally embedded and pursued, which entails 'trying to have a pleasant day and evening and not much more' (Weiss and Bass 2002: 5ff.), whereas in the first setting, prisoners were more reliant on their own resources.

'If all goes well with the men, then I'm also doing well'. How do you know if all goes well? 'Just that you notice that they are comfortable. And, that their day isn't a hell they have to go through. That there is something to do most of the time… that is the most feasible, huh. Mostly, erm, we have to be here, we cannot… we haven't chosen for it, but that it's bearable. That we can get through the day without… having to be melancholic about it all the time'. (Interview Peter, March 2016)

For one of the oldest inhabitants of the specialised unit, a good day equated to one where his sugar level was under control (e.g. Field notes, January 2016). Favouring the prison institute, activities were organised intra-group, whereas some want individually oriented activities or with other groups of prisoners for those housed in the specialised unit, redolent of Goffman's (1961) notion of 'batch living'. In the second prison setting, activities were—albeit not exclusively—aimed at cognitive higher ages. In addition, internalised ageism led to self-censorship amongst some older prisoners: upon asking one of the older inhabitants of the gero-medical unit why he did not want to walk with the other prisoners, he responded: 'What am I still supposed to do over there?' (Interview Antonius, March 2016). This phenomenon was acknowledged by the prisoner-servant in the second setting:

> And a number of other people they do not make use of that offer of activities. They would be able to do it if they wanted to, but there are things which hold them back, uhm, like the feeling of not being able to learn. (Interview Peter March 2016)

From distraction (not thinking about being in prison), time allocation (breaks up the monotony of the day and provides something to look forward to), to physical activity or identity construction (e.g. work), activities perform multiple functions (Meek 2013: 142). The distinction between prison activities that are instrumental (e.g. watching television or reading newspapers to be informed) or otherwise (DeRosia 1998: 98) can also be drawn upon. Reading, in all its manifestations, has already been identified as an inside activity that comprises the bulk of inmates' free time (DeRosia 1998: 97f.). In the course of outlining his day, Ivan described how reading is construed as an attempt to mentally escape problematic situation(s), characterised as a 'lower-level' type of coping response (DeRosia 1998: 97; Zamble and Porporino 1988: 56). Higher-level coping methods are known to involve 'active thinking about one's situation and generally require planning and analysis' (Zamble and Porporino 1988: 60; cf. also DeRosia 1998: 101).

> R: […] And then it's time to read, huh. Then it's time to read. And erm, there are books that are difficult to bear. There are books that are… difficult to read, in my opinion. Such as in *The name of the Rose*, by Umberto Eco. And other books that are in fact too complicated to read in here.
>
> I: What do you exactly mean with 'in here, too complicated to read in here'?
>
> R: By that, I mean that you… if you ponder the whole time, even in your subconsciousness. That you always… even between two lines… You become obstructed. Therefore, it's better to read a thriller that is exciting. And then you continue, and you actually forget your thoughts. (Interview Ivan, November, 2014)

During the course of his confinement, Ivan gradually shifted from reading professional/occupational literature towards fiction. In addition, intellectual stimulation was still sought by solving crossword puzzles ('we still can do something') (Interview, Ivan, November, 2014). This was also a common activity in the sample, which constituted a way of accomplishing goals, challenging oneself, keeping the mind sharp, and maintaining good mental health. This duality is also found in the activity of reading the newspaper: which is a pretext either to keep oneself occupied or to be informed of current events. In similar cases, where prisoners were engaged in constructive activities to 'use' their time, they are regarded to draw upon 'higher-level' coping mechanisms (DeRosia 1998: 97). Most activities that older prisoners were engaged in have been subsumed under what Zamble and Porporino (1988: 96) refer to as 'palliative responses' or 'things that are done to improve prisoners' mood', inter alia, handiwork (such as knitting, crafting, drawing), writing, drawing, watching television, listening to the radio or music, or gaming. By the same token, sleeping or going to bed is described by some older prisoners as a coping mechanism of the inactive kind: 'I feel good now, because I've slept' (Interview Virgil, March, 2016):

> I: Do you sometimes get bored here?
>
> R: No. When I'm not feeling well, I go off to my bed. Like now, I've slept. [...]. I feel good because I slept.
>
> For this prisoner, sleeping was also a form of higher-level coping as 'it is helping him to recover'. (Interview Virgil, March 2016)

Palliative responses, then, are centralised around memories of others or their personal life story; in the absence of people, older prisoners reminisce by looking at photos and letters (cf. Zamble and Porporino 1988). It was observed how one prisoner chose to hang photos of his children on the wall next to his bed, close to him, right next to his head when he slept, and kissed it before bedtime; it provided a perfect vignette of such a response. For older prisoners, little is offered by the prison environment or the regime in terms of higher-level coping strategies, as later life is regarded as sterile—a further example of the detrimental impact of ageism.

When prison activities are geared to low(er)-level coping modes, remediation can only be temporary as it leaves the basic situation unchanged (Zamble and Porporino 1988: 58). Its poor effectiveness in shifting attention is evidenced by Ivan:

> R: I even stopped doing it [solving crossword puzzles] for a while. But uh, now we are eagerly going to continue to do so.
>
> I: Why did you stop?
>
> R: That mental aspect, huh. Because you can... with all that's happening and so forth. A person doesn't forget that easily, I can tell you. If you're somewhat normal, well 'normal', you know what I mean, then you'll...

Think in between the sentences. Then it has no use to do it. And one gets lazy over here. One gets really lazy over here.

I: And why is it that you become so lazy over here?

R: I assume that it's because... you're locked up all the time. Or you go to work, then you work huh. But I can't work because of back complaints. (Interview Ivan, November, 2014)

## Internal Coping

Older prisoners' internal coping, that is to say to resources that exist within the prisoner (Barkway 2013: 240) and require internal reflection (Frydenberg 1997: 57), are intrapsychic mechanisms prisoners employ in addition to, or instead of, those discussed above.

### Social Comparison

The resilience of the 'ageing self' shows that older adults may use intrapsychic ways of coping with unchangeable losses, such as downgrading the importance of lost roles or comparing oneself with others in worse circumstances (Pinquart 2002). Comparing themselves to others who are deemed to be 'worse off' was found to be a common coping strategy amongst the older participants; one participant, for example, engaged in a thinking exercise to remain positive: '[...] remember that there is always someone worse off than you'. The core participants drew on many such empirical instances from other prisoners.

First, in response to stigma, it was found that older prisoners sought 'to regulate the internal psychological effects of their predicament by manipulating relevant social comparisons'. Their frames of reference often pertained to the negative stereotypes of ageing, where they would distance themselves from 'frailer' older prisoners. Moreover, they tended to ignore or redefine concepts to match the definition. This applied particularly to health issues and age-related physical and mental decline. For example, 'Oh, come on, do you see me like people such as Georges walking around with a rollator like that?' (Field notes P1, April, 2014). In other cases, emotional age and maturity were framed. Anita compared herself with another older prisoner performing—at least socially—better in the prison environment whilst altering the comparison: 'I am myself, she is not' (Interview Anita, September 2014). Similarly, one of the older prisoners perceived as 'frail' referred to a younger prisoner who was mute, 'I don't think he can speak. I feel sorry for him. I say to myself "'t is also a poor soul. So young…"' (Interview Antonius, March, 2016).

Secondly, coping strategies were observed that regulated the internal psychological effects of the prison experience. A number of prisoners felt that it was harder for younger persons to be imprisoned than for themselves, as they had less to lose. Although this could be a potential strategy for coping, it could also be regarded as a form of internalised ageism, whereby older prisoners perceive that they have already lived their lives. Conversely, others are more positive about their life stage, e.g. 'I would have had a real good period in life now and I'm stuck here in the joint' (Interview Dorothy, October, 2014), and appraise their current situation as a deprivation.

An overwhelming majority of older prisoners preferred to be housed in prison rather than a nursing home, which strongly counters the preference for older prisoners residing in care facilities. The findings suggest that this does not occur in isolation from functional and/or cognitive and higher social age, where the idea of relocating to a nursing home evokes far less revulsion (Miscellaneous Interviews and Field notes). For most prisoners in the specialised unit, the prospect of relocating to a long-term care facility (for older adults) was received with far less reluctance as it would often be their last resort in terms of leaving prison. It was observed how one such prisoner, revived upon his admission to a nursing home, became 'one of the best', as opposed to one of the 'worst' and most vulnerable residents in the integrated prison.

## Adopting an Attitude of Resignation

Similar to the findings of Reichstadt et al. (2010: 4), a sizable proportion of the older prisoners expressed an attitude of resignation towards their current circumstances: e.g. 'I may not complain' (Interview Antonius, March 2016), whereby the value of not worrying too much about the future was also strongly conveyed.

> I: Does time go slow or fast here?
>
> R: Not slow… Not slow… The days pass well, even the years pass good. Always because it has passed, huh. But I'm never waiting… on this or that. Yeah… We're already advanced this year…. we're almost already New year. Actually, I believe it's… But in my opinion that will have something to do with growing older. 'tis like New year has yet passed. Hm yeah. So, that's not the problem.
>
> R: … But it are all years that I've lost, huh. [sighs]
>
> I: Yes. And how do you cope with that?
>
> R: Gosh, there's nothing you can do about it! Nothing that can be done about it. The only thing we can do is accept it…. (Interview Brian, November, 2014)

A significant number of older prisoners spoke of 'taking every day as it comes'.

> Dorothy also narrates: 'Then it's already half past three and the day is almost over'. She takes it day by day and says at the end: 'One more day is over'. (Field notes, August 2014)

And similarly:

> I take each day as it comes; such as mother serves the soup. Do you understand? (Interview Antonius, March 2016)

This is reminiscent of strategies described by O'Donnell (2014: 227ff.) as 'rescheduling' and 'reorientation', focussing on the present, by breaking it down into manageable compartments.

## Reviewing Life

Aged persons have been described as:

> being in the best position to have evolved a full and satisfying meaning of life, as they have experiences all that the seasons of life have to offer, and because they have achieved most of their important life goals. (Pinquart 2002: 91)

How people look back on their lives is something that might be very disturbing when one is constantly reminded of the reasons behind their imprisonment. This is especially so when there is widespread perception within society that older adults are amongst the most 'law-abiding' (Palmore 1999). Moreover, the possibilities for 'tying any emotional ends' have inevitably become scarcer (Finlay 2009). Accepting and coming to terms with the fact that one's life may end in prison can also become a daunting realisation. Conversely, life review, reminiscence, or nostalgia is something they can (still) draw upon and might just offer a resource as they try to give meaning to their lives. An older prisoner stated: 'I find it harder for the ones who are young in prison… I have already lived a good life. They have to start theirs and are in prison.' (Prisoner, 65+, Field notes P1, December 2014).

## Self-Comparison

At the time of the data collection, those in the sample self-assessed their global subjective well-being within the philosophy of the Anamnestic Comparative Self-Assessment (ACSA),[5] i.e. well-being relative to their memories of the best period and worst period in their lives. Many identified non-prison-related experiences with the worst period(s) in their life. These included illnesses of themselves or significant others; loss of parents; and traumatic experiences such as physical or otherwise abuse. For many their best moments involved joyful youthful experiences, birth of (grand-)children, or meeting their spouse. One individual identified the offence that caused his imprisonment as his worst moment, rather than the experience of imprisonment per se. Conversely, some prisoners indicated that the experience of imprisonment was indeed the worst period in their lives. Unsurprisingly, none suggested that prison was their best moment in life, although several indicated that their current position could be situated between these two extreme periods.

> I: So, in your opinion, these years in prison are the worst in all of your life?
> R: Yes, yes, yes.
> I: And if you compare that to the period you were diagnosed with cancer, and so?
> R: No, you cannot compare that.
> I: It is unlike any other feeling you ever had?
> R: No. It's for real. And that's the way I see that… That's six years of my life that they have stolen from me. Hmm. And truly in hell, huh. […] Note that there are still moments that you have fun, and that you laugh… and that we do stuff… But physically, I'm at least ten years older. And on top of that, always the fear that you have, well, I'm afraid if the cancer will come back, and, you don't have any medical facilities and… (Interview Dorothy, October, 2014)

The (primary) appraisal of prison life as the worst life experiences did not preclude coping. Herein lies the importance of *meaning-based coping* (Folkman 2011), which can mediate the impact of imprisonment. Some older prisoners might be unable to construct an integrated life story, leaving them vulnerable to encountering meaninglessness within a more threatening existential vacuum. The most unfortunate older prisoners viewed their life as an accumulation of negative events with incarceration supplying the final injury. If older prisoners had not yet gained the skills to cope with stressful life events, it is likely that this will continue during imprisonment.

## Cultivating Positive Experiences

In the context of this, older prisoners' reality is not always expressed negatively. Feelings of contentment and enjoyment with life in general have been voiced. Adaptation to prison life was often expressed as 'Making the best of it' (Miscellaneous data excerpts), where prisoners focused on the potential and value in their lives. This prisoner's strategy was aimed at reconceptualising the issue:

> R: It's really… rotated by 360 degrees. [He talks about his current life, relative to pre-imprisonment]
>
> I: And how do you cope with that?
>
> R: You have to. You should not worry too much about it, because that makes it even worse. I always say: I always try to look on the bright side of life [laughs]. And you've probably already noticed that, I'm always… cheerful. Because…If you let your head hang, it doesn't make any difference. You still have to go through it. I have always told John [i.e. psychiatric nurse] from the start. At the beginning, the first few days, they keep an eye on you. Who? The chefs [prison guards]; a matter of suicide. I immediately said to John 'Don't you worry, I love life too much'. You just have to go through it, there's nothing that can be done, you have to make the best of it. (Interview Albert, February 2016)

Moreover, it also appeared to be a matter of not dwelling on negative thoughts and experiences, or as defined by Antonius 'do not dramatise it':

> […] I think a lot about them [i.e. his grandchildren]. But I don't make any drama about it in my head, huh. I take the most beautiful moments that I remember best […]. (Interview Antonius, March 2016)

Taken one step further, some prisoners are driven by 'positive reframing', where they adhere to a constructive interpretation of certain events.

> She says: 'If they talk about me, then I think: "Well, at least I'm still interesting enough to talk about", whilst Anita, if only she has the slightest

suspicion that someone is talking about her, she walks up to the person and asks what they have to say about her.' (Field notes, August 2014)

Events, changes, or challenges were thus interpreted from positive and negative frames. Accordingly, persons trapped within the imprisonment-ageing nexus can view themselves as survivors or as victims (Rybarczyk et al. 2012: 181). Reframing imprisonment in the light of this could be regarded as one of many instances of meaning-based coping. For instance, research into the positive psychological coping techniques of older adults in the wake of illness and disability has provided some evidence to suggest that older adults, as a consequence of their accumulated life experiences and emotional regulation, are in a favourable position to positively reframe events compared to younger adults (Rybarczyk et al. 2012: 181). Can the same be said for older prisoners?

Similar mechanisms were seen to work vis-à-vis objects that were imbued with meaning. For example, a wheelchair, walker, cane/crutch, which for most may ironically constitute 'a mark of dependency', or a perceived impediment to certain activities, such as walking ('I cannot handle this thing properly[Antonius re the walker]', a few viewed these as a symbol of, albeit limited, independence and extended possibilities (Rybarczyk et al. 2012: 181), or something in-between: Albert says 'I have a love-hate relationship with that [wheel]' chair. I ask him to elaborate, whereupon he says: 'Sometimes, I would sometimes want to throw it at the garbage dump, but on the other hand … I would miss it' (Field notes, January 2016).

The narratives of prisoners provide some support for establishing a 'positivity effect', the way 'older adults show more emotionally gratifying memory distortion for past choices and autobiographical information than younger adults do' (Mather and Carstensen 2005). The selectivity in the kind of thoughts about the past is furthermore reconciled with the ability to focus more on the present moment and the preference for short-term emotional well-being (Reed and Carstensen 2012).

I: How do you look back upon what has passed?

R: On the whole, I have had a good life. I don't think about my past anymore, about what I have done wrong.

I: Do you think about the future, or do you live more in the present?
R: I live more in the 'now'; time will tell… Yes. I will put it like this: if I were to start thinking now, and thinking and mulling, thinking, and mulling like 'it's going to be like this', or 'when I get out…', like this or like that, and it turns out differently… Then you're really stuck in the wrong rut, huh. You mustn't think about… 'I would do this and that' Time will tell. (Interview Antonius, March 2016)

According to representatives of this attitude, it is imperative to not allow your thoughts to drift there:

She says that she sometimes also has dark moments, but that you yourself can do a lot. She talks about when she has an off day; then she turns the picture of her parents around, facing away from her, and says out loud: 'I've been a little off today'. She continues: 'And then the next day it's over'. She says she doesn't let herself be 'captured by those moments.' [With her hand she simultaneously makes the gesture that she grabs onto something]. She says she 'always makes sure that she keeps busy', because 'there is a lot of time to think in prison'. Her significant other often states that 'she thinks too much'. (Field notes, September 2014)

Conversely, a very small minority of the core participants seemed to focus on negative information, instead of orienting away from it. Anita and Eugene, for example, were found to focus and dwell particularly on negative experiences, thoughts, and memories (Interview Anita, September, 2014; Interview Eugene, December, 2014).

## Attempts at Mastery

Whilst Crawley and Sparks (2005b: 348, 350) have addressed aspects of aged prisoners' past experiences of coping under the rubric of 'attempts at mastery', Mann (2012: 67) has extended the idea of attempts at mastery to include other previous experiences older prisoners draw upon to survive the rigours of prison. It was also clear in the findings here that many older inmates drew on previous life experiences to cope with imprisonment. These comprised both negative and positive experiences or a combination of both, ranging from previous encounters with institutional life (e.g. in the army, as a captain, at boarding school, sport camp, youth organisation), previous prison experiences, or previous experiences of deprivation (e.g. war, death, loss, illness of the self or significant others). Similar to the prisoner included in Mann's (2012) study, some prisoners drew on previous employment experiences, whereby they had become used to some degree of solitude and being away from family and friends for extended periods of time.

A large number of older prisoners were able to call upon adverse experiences to achieve some sort of 'mastery' to cope with prison life. For prisoners like Antonius, who had spent a part of his childhood at a boarding school run by nuns, some of the deprivations and dehumanising impact of institutional life in a brutalising environment had already been endured earlier in life (Interview Antonius, March, 2016).

R: It's not that bad here. I know places where it's a lot worse.

I: Where?

R: Gosh gosh gosh! … You should stay with nuns. Do you understand me? Yes, with nuns.

I: Have you ever gone there?

R: Yes, three years. In [substituting the name of Flemish city] …. Cowards!

I: How was it over there? Was it tough or hard?

R: Urgh. I hate nuns! [pause] We ate in a large hall… and there was a boy who had an itch. Across his entire body. And he had to stand naked in front of the window and the nun scraped him with ointment and the brush and he wasn't allowed to cry, and we had to laugh. Do I have to tell you more?!

I: How long did that go on?

R: Three years. And I had beautiful hair, curly hair, goddammit, I was the curly-haired kid, huh. And they were going to cut my hair… Everything off! It remained that long [shows].

I: What was your reaction?

R: Angry, huh. Not friendly. And cursing them. There was another nun who lied dead in a shed. And I had to go over there with some other children. You know what I said? 'You smell'. Trust me, I'll never forget it. We had to kiss her, but I didn't do that. I just said: 'You smell'. Nothing more, nothing less. 'You smell!'. That was clear enough, huh?

I: Was that worse than being in prison?

R: With that nun, it was worse. Yes, yes.

I: Did you experience something similar over here?

R: No… no… My mother came to visit me, and she brought candy with her. She gave it to me when she came to visit. But the nuns did not know any better than… Guess what? They took it with them and 'you will get something now and then'. I never saw anything of that candy. Nuns… they are malicious people. They speak of God, and Jesus, well they should better keep their mouth shut.

Several older prisoners had performed (compulsory) military service, which was often mentioned in conversation, e.g.: 'He also refers to his military service and says that this has enabled him to adapt better to prison. He thinks the young people do not know this anymore' (Field notes P2, April 2014). It was striking that many prisoners perceive their time in the military as an experience that has enriched their lives, including both positive and negative aspects. Many of these older prisoners took pride in their status as a soldier and pointed out the differences to their loss of liberty in comparison with prison life.

I: I was wondering if you have earlier experience with another type of institution?

R: Yes, with the army, I've served for two years, so…

I: And is that comparable to life here? Is there something that reminds you of there?

R: Ehm… yes. Yes, yes, over here… that fairly resembles my conscription. Apart from the fact that I'm not allowed to go outside in the evening here [laughs]. After your service, you were normally allowed to go outside, at least if you weren't punished or so on. (Interview Albert, February, 2016)

John, however, pointed to the incomparability between the army and prison, as he was 'healthy' at the time of his military service, thus indicating an importance between dimensions of age and the intra-individual change(s) in recalling a situation as stressful, sometimes in ways that outweighed external environmental factors (Interview John, March, 2016).

Several core participants adopted re-interpretive types of coping such as reevaluation or self-control, (Zamble and Porporino 1988: 57). In such ways, the involuntary aspect to their incarceration, and/or its inherent deprivations (e.g. the loss of freedom), was somewhat mitigated. Although the policy for those residing in individual rooms in the specialised unit at prison setting no. 2 was uniform with regard to the overnight door regime, one core participant strongly asserted the following: 'I ask the guards to lock the door in the evening, because otherwise people always come inside. Then I'm comfortable' (Field notes P2, March, 2016).

Another older prisoner alluded to this type of coping during an observation:

> I ask him about the T-shirt he's wearing. He's wearing a red T-shirt with the acronym of the name of the prison followed by the inscription 'Sport'. He said he was wearing this because this reminded him of a 'sports camp'. And that this enables him to continue his military career. This prisoner was still very active and went to the prison gym 4 times a week despite former cardiovascular interventions. (Field notes P1, April 2014)

These prisoners endeavoured to change the meaning of imprisonment as a stressor without distorting reality (Lazarus and Folkman 1984: 151). For some, that boundary may be crossed, resulting in some form of 'reality distortion'. Evidence was found to suggest some older prisoners' retreat to the past, or to another reality, elicited attempts from prison staff to bring these persons back to 'reality', regarding their feelings as a symptom of a disease that needed to be treated. For instance, one prisoner in the medical unit stated, 'I am in a hospital, not in prison' (quote from Field notes P1, July, 2014); and prisoners in the second setting spoke of 'the hospital' (Miscellaneous Field notes P2). This appeared to be particularly prevalent amongst those with perceived cognitive impairments or deficits (prisoner dubbed as 'the wizard') (Field notes P1, April, 2014).

This is reminiscent of what is known as 'defence mechanisms', which occur without conscious intentionality, effort, and awareness and which contrast with coping mechanisms where there is a conscious, purposeful effort to manage or solve a problem (Cramer 2006: 8). The boundary between the two was at times ambiguous. In contrast to O'Donnell's (Harvey 2012: 62) comment that memory loss can be regarded as either cognitive failure or a coping strategy, it is argued here that disorientation in time or place might also be regarded as a strategy for prisoners in restricting their temporal or spatial focus.

What these coping strategies share is the use of past experiences as frames of reference to reappraise, or to cope with, the stressful and challenging experience of being imprisoned in later life. In this sense, it could be argued that older prisoners have some intrinsic benefits in relying on the increasing 'wealth of experience' associated with growing older (Baars 2012).

## Locus of Control[6] and a Sense of Mastery

Prisoners with a strong(er) internal locus of control attributed their success in coping with prison life more to internal than external factors: 'I think it's largely up to yourself if it doesn't go well here' (Interview Antonius, March 2016). This was also touched upon in the findings:

> Dorothy reflects about Anita [i.e. an older fellow inmate] and tells about the latter that she seeks out loneliness; and that she's both terribly soured and resentful. Dorothy indicates that she herself can somehow relate to it, but still…. (Field notes, September 2014)

For older prisoners, the internal locus of control and a sense of mastery are related to coping. However, for those in highly restrictive conditions, for example, older prisoners of higher age in environments that are ill-suited to their needs, rigidly maintaining an internal locus of control can be highly maladaptive. A manifestation of a rigid internal locus of control can also hinder a person's ability to cope with stressors in relatively disadvantageous environments which elicits existential threats.

## Managing Thoughts and Emotions

In both settings, ageing prisoners usually aimed to cloak expressions of emotions that might indicate vulnerability. However, this was hampered in the second setting due to the open design and communal living. Older prisoners adhered to the male code which prescribed the hiding or denying of emotional displays *en plein public*, to which prisoners are largely socialised (Corley 2001: 106):

> 'No one will see me cry here. I do not cry except for when my father died'. She says: 'I always say: if my father did not bring me to my knees, then nobody can bring me to my knees'. (Interview Dorothy, October, 2014)

This feeds into Greer's observation of emotional coping strategies whereby prisoners tend to manage their experiences rather than sharing them (Greer 2011). There were many analogous narratives across both settings, e.g.: 'I lost trust in people', 'from the moment I entered, I closed my heart' (Interview Dorothy, October, 2014).

> I: Have you been in prison before? In your younger years?
> R: Yes…
> I: Was it heavier now or then?
> R: Now it's harder for me.
> I: Why is that?
> R: Old age huh. Yeah…
> I: What do you mean by that?
> R: I cannot abide anymore as I used to… I cannot hold it back anymore.
> (Interview Leon, March, 2016)

According to this prisoner, it was more difficult to rein in one's emotions with increasing age, which is in stark contrast with what the prisoner culture prescribes. More generally, it has been argued that with ageing there is an increased emotion regulation, although this was found not to be the case in the face of strong events, or when there are high levels of cognitive age putting a strain on emotional age (regulation)

(Scheibe and Carstensen 2010; Urry and Gross 2010), which might also apply for some older persons in view of imprisonment.

## Adopting a Meta-Perspective

A small minority of core participants relied on religion as a means of coping. Some studies show a correlation between religiosity, prisoners coping with the prison experience, and mental health problems (Koenig 1995; Allen et al. 2008). However, the relation between mental health and religiosity or spirituality depends on the individual's sense of abandonment by a higher power (Allen et al. 2008: 696). A minority of core participants actively relied upon religious coping, which was practiced mostly in the privacy of their cells.

The comment below illustrates how Larry, one of the core participants who was beginning a short prison sentence in later life, turned to 'the Lord' to survive his heinous prison experience:

> R: You have to boost your courage a little bit, huh.
> I: Yes. And how do you do that?
> R: By myself, giving strength. And praying…
> I: By praying…?
> R: Yes, yes…
> I: And how else?
> R: I believe… in our Lord. And I cannot help it. He gives me strength. When I'm struggling, I pray. And I thank him at the evening, for the day. (Interview Larry, October 2014)

Some prisoners hinted at adopting a meta-perspective on time and space, whereby they experienced a connection with something bigger than themselves: variant forms of ego-integrity and gerotranscendence. For Antonius, searching for, and looking at, celestial bodies provided him with a meditatory and reminiscing experience:

> R: And now I can see the moon very often. I look if can see the moon somewhere.
> I: Do you see it over here?

R: Well, not always. You almost wouldn't believe it, but if I look through the window here, then it comes out [gestures]. [If this would be the window, then I see it over here sometimes]. And the pole star, not far away from it.

I: You also see that?

R: Yes. And it even happens that I see the pole star on its own. Yesterday evening I saw it and I said to myself 'where is that moon?' But the moon can be over here sometimes, or over there. [...]

I: And from which window do you see that? Do you see it when you're lying in bed or in the room?

R: From my bed ... The pole star, or to the moon.

I: What do you think, when you see that?

R: Well yeah... Pfff... Hhh... A lot of thoughts about the past fly through my head. (Interview Antonius, March 2016)

## The Stories of Older Prisoners

Throughout history, human beings have been described as 'storytelling' animals (MacIntyre 1984: 216). In the narrative gerontological literature, it is recognised that life stories can give meaning to the lives of older adults (Randall 2014). The specific nature of growing older stems from the fact that an important part of life has already taken place (Baars 2012: 217–218). Although the time that has yet to come is in some ways as, or even more, important, the past has already been affected and continues to play a role (Baars 2012: 218). The findings of research on purpose in life amongst the very old showed that very old men experienced purpose in life most strongly when remembering the past and describing their earlier work (Hedberg et al. 2013). Nevertheless, life stories are related to the present, especially with the amount of attention that goes into listening. In addition to participation and involvement in activities, finding meaning has also been linked by penologists to the feeling of 'being listened to' (Liebling 2004: 366). Older adults in prison may use life storytelling or narratives (either oral or written) to give meaning to their life by managing their identity, expressing generativity, or maintaining control over their lives. However, in the absence of an interested audience, narrating could easily become a futile endeavour. The disappearance of one's own narratives, especially when overwritten by prison staff, whose interpretation of their growing older is in terms of their deviant behaviour, is a major threat for meaningfulness in later life. This begs the question as to how prisons could be reconfigured as 'narrative environments' or 'story spaces' (cf. Randall and McKim 2008: 244).

It can be argued that older adult prisoners are offered little from which they can derive meaning by the prison institution. The fact that prisoners, who are spending the later or latest stages of their lives in prison, rely on themselves to make their lives meaningful is exhibited in the section below. Most older prisoners required skills to manage their lives as much as possible during their confinement, with some endeavouring to retain a part of what had been meaningful in their lives in the face of imprisonment and others seeking new ways. A small minority will try to fend off change and might lapse into disengagement and despair (Weiss and Bass 2002).

It can be inferred from the data that older prisoners with unresolved developmental issues, maladjustment, or a weakened social support system were less equipped to cope with imprisonment and the negative interaction with the prison environment. Furthermore, the inmates from the sample extensively drew upon 'escape or avoidance responses', involving staying away from or terminating thoughts about a problem situation (Zamble and Porporino 1988: 94). This is in line with the findings here that some ageing prisoners were disregarded because of ageism, dehumanising processes, prison deprivations, and age-related losses. The findings feed into studies of the 'ageing paradox' (Bonanno et al. 2012: 201; Charles and Carstensen 2010), which alluded to psychological improvements in old age that although older prisoners—in interaction with the dimensions of age—have more to lose, they had proved to be fairly adept in coping.

The majority of core participants were more likely to accept their current situation by virtue of a realistic self-appraisal of their lives and consequently orientated themselves towards the present (including affirmative experiences of the past as being part of the life review) rather than the uncertainty surrounding the future (Reichstadt et al. 2010: 4).

What is denoted as 'coping' in psychology could actually be redefined in terms of 'practical wisdom' (Leijssen 2004) or the 'art of living', practiced by prisoners who fare better during their incarceration (O'Donnell's 2014: 62). This is not associated with chronological age per se but with an attitude that might be considered to fall under the realm of psychological age.

Ian O'Donnell (2014: 62) wrote that his work was 'not intended to downplay the harms of penal isolation but rather to extend the debate by drawing attention to the capacity that some solitaries possess to grow in maturity and wisdom at the same time as they recoil from repeated assaults on their identities' (O'Donnell 2014: 62, own emphasis). Similarly, this chapter should not be perceived as underplaying the adverse impact that imprisonment has on those of old age but as a unique insight into how the human capacity can overcome dire circumstances.

## Notes

1. This author denotes 'older prisoners' on the basis of lower chronological age ($M = 47.46$ yrs, SD = 12.97 yrs) (Vanhooren et al. 2016: 7).
2. The use of the term 'coping' is prompted by the body of existing literature that has employed this term to denote and describe older prisoners' (coping and adapting) *responses* to (respectively, acute and chronic) stressors of imprisonment in an endeavour to retain meaning (Mann 2012: 63ff.; Crawley and Sparks 2005b: 343ff.).
3. Whilst Sykes (1958/2007) has already hinted at the differences between individuals in experiencing these pains of imprisonment, the different dimensions of (old) 'age' have been rarely explored in view of this.
4. Relations with fellow prisoners have been subsumed either under the heading of the social prison environment (see Chap. 4) or under the concept of generativity (cf. Chap. 6).
5. ACSA has been introduced by Bernheim and his colleagues (2006) to measure the subjective well-being of patients. Rather than asking participants to rate their well-being on the anchored ASCA-scale, we merely employed the underlying philosophy of ACSA by asking them about their

worst and best period in their life, and, how they would position themselves at the moment in comparison to these periods.
6. We have opted to use the term 'locus of control' rather than 'attribution of responsibility' (Bors and Stevens 2010: 57ff.), as it has been frequently used in the literature on (older) prisoners (cf. Aday who argued that locus of control becomes more important with increasing age (2003: 150)).

## References

Aday, R. H. (2003). *Aging Prisoners: Crisis in American Corrections*. Greenwood Publishing Group.

Allen, R. S., Phillips, L. L., Roff, L. L., Cavanaugh, R., & Day, L. (2008). Religiousness/Spirituality and Mental Health Among Older Male Inmates. *The Gerontologist, 48*(5), 692–697. https://doi.org/10.1093/geront/48.5.692.

Baars, J. (2012). *Het nieuwe ouder worden. Paradoxen en perspectieven van leven in de tijd* (3rd Edition). SWP Uitgeverij.

Barkway, P. (2013). Stress and Coping. In P. Barkway (Ed.), *Psychology for Health Professionals* (pp. 222–250). Elsevier Health Sciences.

Baumeister, B., & Keller, S. (2011). *Alt werden im Straf- und Massnahmenvollzug*. ZHAW Soziale Arbeit.

Bernheim, J. L., Theuns, P., Mazaheri, M., Hofmans, J., Fliege, H., & Rose, M. (2006). The Potential of Anamnestic Comparative Self-Assessment (ACSA) to Reduce Bias in the Measurement of Subjective Well-Being. *Journal of Happiness Studies, 7*(2), 227–250. https://doi.org/10.1007/s10902-005-4755-0.

Bonanno, G. A., Westphal, M., & Mancini, A. D. (2012). Loss, Trama, and Resilience in Adulthood. In B. Hayslip Jr. & G. C. Smith (Eds.), *Annual Review of Gerontology and Geriatrics, Volume 32, 2012: Emerging Perspectives on Resilience in Adulthood and Later Life* (pp. 189–210). Springer Publishing Company.

Bond, G. D., Thompson, L. A., & Malloy, D. M. (2005). Lifespan Differences in the Social Networks of Prison Inmates. *International Journal of Aging & Human Development, 61*(3), 161–178. https://doi.org/10.2190/7H5P-2AHJ-L34Q-GW9U.

Bors, G., & Stevens, L. (2010). *De gemotiveerde leerling*. Garant.

Carlisle, D. (2006). So far, so bleak: Increasing numbers of older prisoners in a prison estate, designed essentially for fit young men, pose a problem for health service providers. *Nursing Older People, 18*(7), 20–23.

Carstensen, L. L. (1992). Social and Emotional Patterns in Adulthood: Support for Socioemotional Selectivity Theory. *Psychology and Aging, 7*(3), 331–338.

Charles, S., & Carstensen, L. L. (2010). Social and Emotional Aging. *Annual Review of Psychology, 61*, 383–409. https://doi.org/10.1146/annurev.psych.093008.100448.

Cohen, S., & Taylor, L. (1972). *Psychological Survival: The Experience of Long-term Imprisonment*. Penguin Books.

Corley, D. (2001). Prison Friendships. In D. F. Sabo, T. A. Kupers, & W. J. London (Eds.), *Prison Masculinities* (pp. 106–108). Temple University Press.

Cramer, P. (2006). *Protecting the Self: Defense Mechanisms in Action*. Guilford Press.

Crawley, E. (2005). Institutional Thoughtlessness in Prisons and its Impacts on the Day-to-Day Lives of Elderly Men. *Journal of Contemporary Justice, 21*(4), 350–363.

Crawley, E., & Sparks, R. (2005a). Hidden Injuries? Researching the Experiences of Older Men in English Prisons. *The Howard Journal, 44*(4), 345–356.

Crawley, E., & Sparks, R. (2005b). Older Men in Prison: Survival, Coping and Identity. In A. Liebling & S. Maruna (Eds.), *The Effects of Imprisonment* (pp. 343–365). Willan Publishing.

DeRosia, V. R. (1998). *Living Inside Prison Walls: Adjustment Behavior*. Greenwood Publishing Group.

Farber, M. L. (1944). Suffering and Time Perspective of the Prisoner. In K. Lewin (Ed.), *Authority and Frustration*. University of Iowa.

Finlay, H. (2009). Older men in prison: Making sense of aging. *Prison Service Journal, 185*, 28–34.

Folkman, S. (2011). *The Oxford Handbook of Stress, Health, and Coping*. Oxford University Press.

Frankl, V. E. (1959). *Man's Search for Meaning*. Pocket Books.

Frydenberg, E. (1997). *Adolescent Coping: Theoretical and Research Perspectives*. Psychology Press.

Goffman, E. (1961). Asylums: Essays on the social situation of mental patients and other inmates. Anchor Books.

Greer, K. (2011). Walking an Emotional Tightrope: Managing Emotions in a Women's Prison. *Symbolic Interaction, 25*(1), 117–139. https://doi.org/10.1525/si.2002.25.1.117.

Harvey, J. (2012). *Young Men in Prison: Surviving and Adapting to Life Inside*. Routledge.

Hedberg, P., Gustafson, Y., Brulin, C., & Alex, L. (2013). Purpose in life among very old men. *Advances in Aging Research, 2*(3), 100–105. https://doi.org/10.4236/aar.2013.23014.

Humblet, D., & Decorte, T. (2013). Detentiebeleving door oudere onderzoek-gedetineerden in België: Een exploratief. Panopticon. *Tijdschrift voor strafrecht, criminologie en forensisch welzijnswerk, 34*(4), 267–283.

Koenig, H. G. (1995). Religion and Older Men in Prison. *International Journal of Geriatric Psychiatry, 10*(3), 219–230. https://doi.org/10.1002/gps.930100308.

Krabill, J. J., & Aday, R. H. (2007). Exploring the Social World of Aging Female Prisoners. *Women & Criminal Justice, 17*(1), 27–53. https://doi.org/10.1300/J012v17n01_02.

Kuin, Y., & Westerhof, G. J. (2007). Zingeving. In A. M. Pot, Y. Kuin, & M. T. Vink (Eds.), *Handboek ouderenpsychologie* (pp. 175–190). Tijdstroom.

Laws, B. (2016). *Fronting, Masking and Emotion Release: An Exploration of Prisoners' Emotional Management Strategies.* The Howard League for Penal Reform.

Lazarus, R. S. L., & Folkman, S. F. (1984). *Stress, Appraisal, and Coping.* Springer Publishing Company.

Leigey, M. E., & Ryder, M. A. (2015). The Pains of Permanent Imprisonment: Examining Perceptions of Confinement Among Older Life Without Parole Inmates. *International Journal of Offender Therapy and Comparative Criminology, 59*(7), 726–742. https://doi.org/10.1177/0306624X13517868.

Leijssen, M. (2004). Zingeving en zingevingproblemen vanuit psychologisch perspectief. In Wetenschappelijk denken: Een laboratorium voor morgen? Lessen voor de eenentwintigste eeuw. In B. Raymaekers & G. Van Riel (Eds.), *Wetenschappelijk denken: Een laboratorium voor morgen? Lessen voor de eenentwintigste eeuw* (pp. 241–268). Universitaire Pers Leuven.

Liebling, A. (2004). *Prisons and Their Moral Performance: A Study of Values, Quality, and Prison Life.* Oxford University Press.

Lyon, B. L. (2012). Stress, Coping, and Health: A Conceptual Overview (Update). In V. H. Rice (Ed.), *Handbook of Stress, Coping, and Health: Implications for Nursing Research, Theory, and Practice* (2nd ed., pp. 2–21). SAGE Publications.

MacIntyre, A. (1984). *After virtue: A study in moral theory.* University of Notre Dame Press.

Mann, N. (2008). *Doing harder time? The experience of an ageing male prison population in England and Wales.* University of Essex.

Mann, N. (2012). *Doing Harder Time? The Experiences of an Ageing Male Prison Population in England and Wales.* Ashgate.

Maruna, S., Wilson, L., & Curran, K. (2006). Why God is Often Found Behind Bars: Prison Conversions and the Crisis of Self-Narrative. *Research in Human Development, 3*(2–3), 161–184. https://doi.org/10.1080/15427609.2006.9683367.

Massoglia, M., & Pridemore, W. A. (2015). Incarceration and Health. *Annual Review of Sociology, 41*(1), 291–310. https://doi.org/10.1146/annurev-soc-073014-112326.

Mather, M., & Carstensen, L. L. (2005). Aging and Motivated Cognition: The Positivity Effect in Attention and Memory. *Trends in Cognitive Sciences, 9*(10), 496–502. https://doi.org/10.1016/j.tics.2005.08.005.

Meek, R. (2013). *Sport in Prison: Exploring the Role of Physical Activity in Correctional Settings*. Routledge.

Moody, H. R. (2002). The Changing Meaning of Aging. In R. S. S. Weiss & S. A. Bass (Eds.), *Challenges of the Third Age: Meaning and Purpose in Later Life* (pp. 41–54). Oxford University Press.

Moran, D. (2012). Prisoner Reintegration and the Stigma of Prison Time Inscribed on the Body. *Punishment and Society, 14*(5), 564–583. https://doi.org/10.1177/1462474512464008.

O'Donnell, I. (2014). *Prisoners, Solitude, and Time*. Oxford University Press.

Palmore, E. (1999). *Ageism: Negative and Positive* (2nd ed.). Springer Publishing Company.

Pierce, J., & Shanan, A. (2017). Quality of Life in the Animal Hospice and Palliative Care Patient. In A. Shanan, J. Pierce, & T. Shearer (Eds.), *Hospice and Palliative Care for Companion Animals: Principles and Practice* (pp. 27–40). John Wiley & Sons.

Pinquart, M. (2002). Creating and Maintaining Purpose in Life in Old Age: A Meta-analysis. *Ageing International, 27*(2), 90–114. https://doi.org/10.1007/s12126-002-1004-2.

Prandi, J. D. (1993). *'Dare to be Happy!': A Study of Goethe's Ethics*. University Press of America.

Randall, W. L. (2014). *The Stories We Are: An Essay on Self-Creation* (2nd ed.). University of Toronto Press.

Randall, W. L., & McKim, E. (2008). *Reading Our Lives: The Poetics of Growing Old*. Oxford University Press.

Reed, A. E., & Carstensen, L. L. (2012). The Theory Behind the Age-Related Positivity Effect. *Frontiers in Psychology, 3*, 339. https://doi.org/10.3389/fpsyg.2012.00339.

Reichstadt, J., Sengupta, G., Depp, C. A., Palinkas, L. A., & Jeste, D. V. (2010). Older Adults' Perspectives on Successful Aging: Qualitative Interviews. *The American Journal of Geriatric Psychiatry: Official Journal of the American Association for Geriatric Psychiatry, 18*(7), 567–575.

Rybarczyk, B., Emery, E. E., Guequierre, L. L., Shamaskin, A., & Behel, J. (2012). The Role of Resilience in Chronic Illness and Disability in Older Adults. In B. Hayslip Jr. & G. C. Smith (Eds.), *Annual Review of Gerontology and Geriatrics, Volume 32, 2012: Emerging Perspectives on Resilience in Adulthood and Later Life*. Springer Publishing Company.

Scheibe, S., & Carstensen, L. L. (2010). Emotional Aging: Recent Findings and Future Trends. *The Journals of Gerontology: Series B, 65B*(2), 135–144. https://doi.org/10.1093/geronb/gbp132.

Schmid, W. (2015). *Gelatenheid. Wat we winnen wanneer we ouder worden*. De Bezige Bij.

Segerstrom, S. C. (2006). *The Glass Half Full: How Optimists Get What They Want from Life - and Pessimists can Too*. Constable& Robinson Ltd.

Suurmond, J.-J. (2013). *Meer geluk dan grijsheid: Spiritualiteit van de ouderdom*. Boekencentrum.

Sykes, G. M. (1958). *The Society of Captives*. Princeton University Press.

Sykes, G. M. (2007). *The Society of Captives: A Study of a Maximum Security Prison* [1st Princeton classic ed.]. Princeton University Press.

Urry, H. L., & Gross, J. J. (2010). Emotion Regulation in Older Age. *Psychological Science, 19*(6), 352–357.

Vandebosch, H. (2004). Media Use as an Adaptation or Coping Tool in Prison. In K. Renckstorf, D. McQuail, J. E. Rosenbaum, & G. Schaap (Eds.), *Action Theory and Communication Research: Recent Developments in Europe. (MoutonTextbook)* (2004 ste dr. ed., pp. 199–216). Mouton de Gruyter.

Vanhooren, S., Leijssen, M., & Dezutter, J. (2016). Profiles of Meaning and Search for Meaning Among Prisoners. *The Journal of Positive Psychology, 11*(6), 622–633. https://doi.org/10.1080/17439760.2015.1137625.

Wahidin, A. (2002). Reconfiguring Older Bodies in the Prison Time Machine. *Journal of Aging and Identity, 7*, 177–193. https://doi.org/10.1023/A:1019764506372.

Wahidin, A. (2004). Older Women in the Criminal Justice System: Running Out of Time. In *Jessica Kingsley Publishers*.

Wahidin, A., & Tate, S. (2005). Prison (E)scapes and Body Tropes: Older Women in the Prison Time Machine. *Body & Society, 11*(2), 59–79. https://doi.org/10.1177/1357034X05052462.

Weiss, R. S., & Bass, S. A. (2002). Introduction. In R. S. S. Weiss & S. A. Bass (Eds.), *Challenges of the Third Age: Meaning and Purpose in Later Life* (pp. 3–12). Oxford University Press.

Willis, A. K., & Zaitzow, B. H. (2015). Doing "Life": A Glimpse into the Long-Term Incarceration Experience. *Laws, 4*(3), 559–578. https://doi.org/10.3390/laws4030559.

Yorston, G. A., & Taylor, P. J. (2006). Commentary: Older Offenders—No Place to Go? *Journal of the American Academy of Psychiatry and the Law Online, 34*(3), 333–337.

Zamble, E., & Porporino, F. J. (1988). *Coping, Behavior, and Adaptation in Prison Inmates*. Springer-Verlag.

# 6

# 'Thriving' in Prison in Old Age

This chapter is concerned with how generativity enables older prisoners to thrive in the prison world, which takes them beyond mere surviving. It is argued here that generativity plays a hitherto underexplored role in explaining the ways in which older prisoners develop meaning. To date, the concept of generativity amongst prisoners in later life has only received very little scholarly scrutiny. The concept of generativity has only fairly recently been introduced in—primarily developmental and, less directly, ecological—criminology and has been in particular linked to criminal behaviour and to the process of desistance from offending (Halsey and Harris 2011; Maruna et al. 2004). To our knowledge, still very little empirical work on imprisonment has been generated by Erikson's concept of generativity. In our research, generativity emerged as an important potential source of meaning in life. Building on the contributions of previous scholars including de St Aubin (2013) and Miranda-Chan and Nakamura (2015), who argue that generativity is *one* potential path to the experience of life meaning, this chapter examines the meaningfulness of generativity for individuals who are spending the later stages of their lives in prison.



D. Humblet, *The Older Prisoner*, Palgrave Studies in Prisons and Penology,
https://doi.org/10.1007/978-3-030-60120-1_6

## Explaining Generativity

The term generativity was coined by the German psychoanalyst and former disciple of Freud Erik Erikson to primarily denote 'the concern in establishing and guiding the next generation [...]' (Erikson 1950: 240). Although Jung did not use the term generativity in his theory of 'individuation',[1] he had touched upon the interest, or wish, in leaving something behind during the second half of life. Erikson, however, used generativity to denote a specific developmental stage in life: the seventh of eight stages (Erikson 1950). In Erikson's framework, each stage is characterised by a conflict (also termed 'task' or 'crisis') that must be resolved and which the outcome is dependent on the outcome of conflict in previous stages, a predetermined outcome Erikson termed the 'epigenetic principle'. According to his theory, the most important task in mid-to-later adulthood is to develop a sense of altruism towards other people and to the next generation. He initially stated that the concept of generativity was typically achieved through biological parenthood, albeit tempered by creativity and productivity:

> [...] there are individuals who, through misfortune or because of special and genuine gifts in other directions, do not apply this drive to their own offspring. And indeed, the concept generativity is meant to include such more popular synonyms as productivity and creativity, which, however, cannot replace it. (Erikson 1950: 240, original emphasis)

However, it was not until the 1980s that his concept of generativity caught the attention of the scientific community as a topic for empirical investigation (McAdams and Logan 2004: 15). An emerging body of research has expressed scepticism about Erikson's very strong emphasis on the care of one's progeny in determining generative needs. More recent work has enlarged the Eriksonian approach by defining generativity as follows: 'a desire to invest one's substance in forms of life and work that outlive the self' (Kotre 1984: 10). Kotre (1984) distinguished four types of generativity: biological (reproduction), parental (caring for children), technical (teaching skills), and cultural (passing on meaning systems) generativity (Kotre 1984, 1996: 10-16). Similarly, in his study of Erikson's generativity concept amongst middle-aged men (1978), and later amongst women (1996), Levinson (1978) pinpointed the importance of 'mentoring' in adult development and suggested that mentor relationships may be equal, or more significant, to achieving generativity than caring for the offspring. In his later writings, such as *Gandhi's Truth*, Erikson (1993) showed how generativity may be expressed outside of the private realm of the family, in public political action through leadership (Erikson 1993; McAdams and Logan 2004: 22). By applying the idea of generativity to older residents at a long-term dementia care setting, Doyle et al. (2015) further refined and expanded the traditional conceptualisation of generativity. Rather than reducing generativity to activities that are solely oriented towards younger or future generations or that must 'outlive the self', generativity may be regarded as 'any sort of activity or intention that may be beneficial to others through the investment of one's self' (Doyle et al. 2015: 2). From this point of view, generative activities or intentions may even range from small things and can thus involve people in the same or in previous generations (Doyle et al. 2015: 2). Our observations accord with these authors' contention that people, groups, objects, and activities can be all viewed as potential vectors for generativity (Doyle et al. 2015: 4).

When considering the most common understandings of 'generativity', it is striking how far it embodies elements of dependency and reciprocity; in a sense, it encompasses 'the need to be needed' by others, regardless of whether this need pertains to intra- or inter-generational counterparts and/or other entities. Several theorists (McAdams 1988: 73 et seq.; Kotre 1996: 16-18)[2] have previously drawn on the work of Bakan (1966) to demonstrate the *dual nature* of generativity, as something 'selfish' and 'agency' oriented on the one hand, and/or as an act of 'selflessness' a 'communal' oriented component on the other. Bakan (1966) suggested that agency places emphasis on the individual striving to master the environment, to assert the self, to experience competence, status, victory, achievement or responsibility, and power. Communion, by contrast, places emphasis on participation in a larger organism of which the individual is part of and refers to the desire to closely relate to, cooperate, and merge with others (Bakan 1966: 14-15; McAdams 2001).

## The Value of Generativity

More than half a decade ago, Erik H. Erikson published his seminal framework of human psychological development wherein he delineates eight stages from 'infancy' to 'old age' (Erikson 1950). Whilst Erikson's initial idea of the life cycle only comprised seven stages, in the prelude to the publication of *Childhood and Society*, a new seventh stage entitled 'generativity vs. stagnation' in adulthood was added, followed by old age with the strengths of wisdom and integrity soon promoted to the eighth stage (Erikson and Erikson 1997). In his first book, Erikson places a very strong emphasis on the childhood stages, which could account for the fact that the concept of generativity received little attention at that time (Erikson 1950: 240). Moreover, as Erik Erikson and his wife Joan Erikson grew older themselves, these authors began to acknowledge the importance of generativity in old age and had reason to believe it had a more important role to play in later life than they initially assumed (Erikson and Erikson 1997):

> Returning once more to the upper right corner, we retrace one step back along the diagonal only to re-enter the *generative* stage that preceded old age. But in an epigenetic scheme, we said, 'after' should mean only a later version of a previous item, not a loss of it. And indeed, old people can and need to maintain a *grand*-generative function. For there can be little doubt that today the discontinuity of family life as a result of dis-location contributes greatly to the lack in old age of that minimum of vital involvement that is necessary for staying really alive. And lack of vital involvement often seems to be the nostalgic theme hidden in the overt symptoms that bring old people to psychotherapy. Much of their [older people's] despair is, in fact, a continuing sense of stagnation. (Erikson and Erikson 1997: 63, original emphasis)

There is some consensus that individuals can be generative throughout their whole life (see, e.g. Kotre 1996), whilst Erikson links this to a certain life phase (cf. supra). In addition, there is some research to suggest that levels of generative interests increase with age (McAdams & de St. Aubin, 1992), though evidence towards an age-related increase in levels of generativity is mainly mixed to date (Maxfield et al. 2014). Whilst some research suggests that personal goals become increasingly generativity-oriented as one ages, others have put forward that younger and older adults do not differ in reported generative concern (see for an overview of studies Maxfield et al. 2014). Accounting for this discrepancy, Maxfield et al. (2014) found that age differences only emerged under conditions of existential threat, suggesting that although people of many ages show some interest in generativity, it is primarily older adults who use it as a source of protection from death-related concerns (cf. de St Aubin). Whichever view be held is dependent upon whether, or to which extent, generativity relates to the finitude of life. Prompted by her own ageing experience, Joan Erikson added another stage, i.e. the ninth stage[3], wherein all the eight stages are relevant and recurring, albeit with the dystonic elements of each stage first, prompted by the circumstance of 'old age'. As to the stage of generativity, she wrote:

> Generativity, which comprised the major life involvement of active individuals, is no longer necessarily expected in old age. This releases elders from the assignment of caretaking. However, not being needed may be felt as a designation of uselessness. When no challenges are offered, a sense of stagnation may well take over. Others, of course, may welcome this as a promise of respite, but if one should withdraw altogether from generativity, from creativity, from caring for and with others entirely, that would be worse than death. (Chapter 5, Erikson and Erikson 1998: 112)

Benedict (2009) related the idea of generativity to incarceration and pointed to the authoritative nature of the prison institution where inmates are separated from free society. As a result, incarcerated adults might be unable to express generativity in ways similar to community-dwelling individuals or face institutional barriers to being generative (Benedict 2009: 51). In her prison study of older women, Benedict stipulated generativity as 'the desire to give back and execution of that desire' (Benedict 2009: 51).

The small body of research on old-age generativity in custodial environments does not correlate generativity with subsequent offending (cf. Maruna 2001). However, there are links between generativity and peaceable behaviour in prison, when generativity-minded older inmates seek out helping, rather than harming, opportunities behind bars. Benedict (2009: 152), for example, suggested that generativity was found amongst older women prisoners, but her methodological approach, i.e. using convenience sampling based on self-selection and the intervention of a prison staff member, raised issues of sample selection bias. More significantly, the 'True Grit program'[4], Kopera-Fry et al. (2012), used data collected by a mixed-method survey amongst older male adult inmates (50+). They looked at the qualitatively obtained generativity-oriented responses as part of a larger pool of positive affective responses, whereby negative affective responses were coupled to the stress of the (prison) environment and the questioning of ego integrity. Based on their analysis of 171 individual responses to the open-ended question examining what it is like being an older adult in prison, they arrive at the following conclusion:

> What is evident [...] is a very stressful environment coupled with a questioning of one's own ego integrity (e.g., 'embarrassed' came up repeatedly) and a feeling of missing out on life's milestones, e.g., retirement. However, the resilience of these men in coping with a stressful context is evident in their positive responses noting that they help their fellow inmates and care deeply about helping younger adults, e.g., generativity. (Kopera-Fry et al. 2012: 50)

In the light of these specific studies, this research fills an evident gap in the extant literature on generativity by employing both interviews and observations with older adults in a peculiar type of total institution.

## Generative Behaviour Amongst Older Prisoners in the Two Settings

### The Generative Motivation to Participate in the Research

For some participants, the taking part in the study reflected a generative rationale. Although some participants, whilst reflecting on the impact of my research, clearly stated that any transformation of the current prison system would '[…] probably not apply to us anymore (laughs)' (Field notes, July, 2016), they were nevertheless hoping that their contributions would be beneficial for future generations of inmates. Whilst very few individuals refused to participate in the research interviews, this was partially motivated by their view that they had nothing of significance to disclose, rather than objections to the research per se.

Upon leaving the research setting(s), most individuals who were closely involved were keen to speculate on the future impact of the research, on any social valorisation of the result, and were very eager to see the final study. In this sense, they clearly perceived being part of a larger endeavour and further that they were contributing to a higher cause, a larger whole, within which they were now embedded. Similarly, protective and patriarchal behaviour towards a 'younger' female researcher was apparent. Some even explicitly stated that I strongly reminded them of their own children or their own grandchildren, even to the extent of being wrongfully mistaken as 'the daughter of…'. To some extent, this attributed to interactions where participants offered advice and guidance by sharing their knowledge gained by life experience, such as the following tip: "If you are alone at sea and you are lost, you should sail to the south. Never forget it!! It may come in handy for you some day, you never know" (Antonius, prisoner, 79, male). Mirroring the generative behaviour of one of Doyle et al. (2015) residents, this participant provided similar advice for getting through life. This included, amongst other things, their views on love and marriage, health, career, and/or legal advice. By telling their life stories, they were not merely (re-)constructing their identities but also passing on their knowledge and values to the next generation, in this case, a researcher, as a guide to avoid repeating their previous mistakes.

## Finding or Taking on Generative Roles in Prison

### Relationship with Peers

Alongside prisoners' generative attitudes towards me, opportunities for some to find generative roles or carry out generative developmental tasks within the prison community were witnessed in various manners and contexts: for example, the advice-giver; the caregiver or caretaker; and the experienced and lived through person, with each showing similarities to the role of a (grand)parent. In most cases, this was performed by an ageing prisoner vis-à-vis a 'younger' counterpart(s).

According to a staff member, the presence of older women who took on the role of grandmother in the integrated prison had a positive and calming effect on the other prisoners (Field notes, June,2014). The (grand)mother role of one female is clearly illustrated by her nickname of 'Mummy' or 'Bonnie' (Field notes, August 2014). Her generative acts manifested themselves through, inter alia, baking and cooking dinner for other prisoners (e.g. making pancakes for the rest); navigating others through daily prison life (e.g. translating or clarifying prison rules, offering services); and offering emotional support and solace. This is amplified in the field notes:

> She also ponders whether it would be possible to organise an intergenerational activity within prison such as babysitting on the infants or children who are staying with their mothers in prison at the prison mother-child department. In the same breath, she suggests the possibility to support each other by visiting each other in cell (even for 15') or even exceptionally sleeping with another prisoner in the same cell when she is going through a difficult time. (Field notes, September 2014)

This would suggest that the type of prison regime, and especially the level of enclosure (open or closed prison) and seclusion (age segregation or seclusion), can have an impact on the opportunity to act upon and shape generative behaviour of this nature.

Another female prisoner (aged mid-60) was also hailed as an 'elderly figure'. In the prison system, she was highly regarded by prisoners and even by staff, mainly because of her prison experience (prison recidivist), attitude, connections to the outside world, and her category of offence. In the prison staff member's view, these two participants competing for a similar role sometimes lead to difficulties (Field notes, June 2014). However, for the oldest female prisoner on this wing (Anita, age 69), it was not possible to interact with younger prisoners in the same generative way as her peers. For her, ageing meant 'maturing', leaving little opportunity to interact with younger prisoners, thus cutting off from them. This could be attributed to advanced biological ageing and socio-psychological ageing, as she encountered difficulties in communicating and connecting with other, mainly chronologically much younger, prisoners due to a hearing impairment and a lack of common interests and conversation topics attributed by herself to an intergenerational gap. Taking into account her background as a widow advanced in age who was involuntarily childless and had virtually no next of kin, the closest Anita came to achieving generativity through intergenerational contact was experiencing the stories and pictures of the children of volunteers who infrequently visited her. The pictures of these children on the blackboard in her cell and the dismissive behaviour towards her experience with miscarriage revealed not only her desire for intergenerational generativity but also its painful absence.

Conversely, in one case, a younger prisoner, who held the position of a prisoner-servant, in his late 20s to early 30s, explicitly stated that he viewed one of the oldest prisoners 'as if it were his own grandfather' (Field notes P1, May 2014). He provided help and support for this older with, for example, cell cleaning, although the older prisoner remained passive in his generative development. This view was more implicitly echoed by another prisoner (also servant, in his mid-50s), who hoped that others would have the same respect for him when he grew older (Interview Peter, March 2016). It could therefore be argued that taking the narrow definition or merely 'generativity' in the broadest sense, these younger prisoners were in fact demonstrating an 'inverted generativity' in their actions.

Peter also stressed a further significance to the idea and importance of generativity through intergenerational relationships:

> Well, this is in itself… it is a part of life and uhm, if there are no counter-indications, then you mustn't… Well, I'm always like: there are many possible variations of normal life. I mean… I have always enjoyed watching other cultures, and they sometimes live so different from us. So, you can do things entirely different, but eventually there are certain things, those connections between generations. […] In our culture, people are easily locked up in their own generation. You must be extremely vigilant not to lose contact with other generations. And that's an impoverishment. That is solely impoverishment. […] Take for instance the examples from classical antiquity, where you had to have relationships that include a vast range of generations. That ensures the strengthening of the cohesion in society. Older people will then not grow lonely, it's as simple as that. […] Well, you can also see that it is becoming almost a necessity, because nowadays there are kindergartens who are going to play at nursing homes. The toddlers are going to play with the elderly, even with demented. That says in every way… those toddlers become calmer as a result, hey, and they also learn to associate with an older person, and they even enjoy it. Because they have time and patience [laughs] and they tell weird things. But I mean… It is not all that difficult. And what we do is just always pigeonholing.
>
> I: Bearing that in mind, would you advance the idea of integrating older adults with younger adults in prison? Or do you think that is a different story?
>
> R: During certain activities, where it is possible… and what they indeed do is those leisure activities and courses, there that is the case… Yes. Whilst of course, well, I reckon… Naturally, that small child does not sleep in the bedroom of the grandpa and grandma, huh? And there is also… the hours that they have a rest and the hours that they go to sleep could perhaps even be comparable, but the noise they make is different: the grandfather snores [laughs] and the new-born cries. (Interview Peter, March, 2016)

His responses strongly resonate with the call for integrating older prisoners amongst younger prisoners, albeit with a specific approach to their needs. Whereas an emerging body of literature is focussing on the differences between groups of prisoners (mostly on the basis of their chronological, psychological, and/or biological age), penitentiary settings are mainly focussing on one specific similarity, i.e. the mere status as a 'prisoner' (cf. sameness principle). Although a mix of different generations is in itself not sufficient to ensure generativity amongst generations of prisoners (cf. the account of Anita, age 69), it is clear that a rigorous age-segregated prison policy hampers or even excludes certain opportunities to engage in intergenerational generativity amongst the prisoner population by default.

## Intragenerational Relationships

Albert (73 years of age) amplified his experiences in the interview:

> 'My colleague is at the moment at another prison, which now makes me the oldest one.' The researcher asks him: 'Your colleague?' to which he responds: 'Well, yeah, "colleague"…'. [He was in fact referring to a former fellow inmate]. He subsequently describes him as: 'The one with the cramped fingers, he was here last time'. The researcher poses the following question: 'And do you mind, being the oldest one now?' He replies: 'Well…, no, but it's the thought of being the oldest one, huh.' Further elaborating on his role, he states that he has noticed that 'they' [coprisoners] sometimes come to ask him about things. He also says that there once was someone who played with the idea of suicide. Albert said that he was able to change this person's mind by talking with him, and that this made him feel good about himself. (Field notes P2, January, 2016)

Later, in his interview, he added that he befriended a former fellow prisoner, whom he perceived to be in his late 40s/early 50s and provided him with good advice to quit smoking, and spend more time on the telephone by purchasing a larger amount of call credit, enabling conversations with a new partner who he met whilst being imprisoned (Interview Albert, February, 2016).

When the psychiatric nurse plays specialised games with older prisoners designed to benefit those with memory problems (Moll 2013), Albert says that he only participates in these games when there is a shortage of players. When asked about his rationale for this, he replies quietly: 'I can't say this now…'. Albert waited for the other participants to be called away before adding in a whisper: 'those games are actually for the weaker ones; I sometimes join them because I'm someone who doesn't mind to lose deliberately' (Field notes, January, 2016). Following Doyle et al. (2015: 2), generativity is expressed here as an activity, or an intention, that may be beneficial to others through the investment of one's self.

## Relationship with Prison Staff as Dyadic Relationships

Small acts of generosity vis-à-vis the prison staff were encountered. However, these were fairly limited and were kept to an absolute minimum, as well as often initiated by the officers themselves ('reactive'): collecting soda pop tabs with a view to recycle it in craftwork; utilising their talents (e.g. drawing a picture on demand); and providing a listening ear or advice regarding staff's personal problems (e.g. relationship issues). These dyadic relationships could therefore be regarded as unilateral and asymmetric. On the one hand then, generativity, by its very nature, has the potential to bridge the divide between staff and inhabitants, yet on the other, the staff-inmate relationship does seemingly not constitute the most robust foundation to accomplish generative desires, as it lacks the trust and reciprocity essential for generativity. Moreover, some prisoners have openly questioned the legitimacy of the prison officer as a profession. To some extent, there is some interconnectedness and sameness amongst prisoners ('we are all in the same boat') that is lacking in the relationship with prison officers. This is consistent with the small body of literature on generativity within custodial institutions where prisons are depicted as environments wherein opportunities to care in non-violent and durable ways for self, other, and future prisoners are reduced to a focus on staff-prisoner relationships:

> In the main, and acknowledging that one prison can never be precisely like another, such environments tend—either by bureaucratic or sub-cultural design—to be places where displays of genuine concern and respect for others (especially the kind which bridges the staff/prisoner divide), or indications of the desire to shape one's future in ways disconnected from crime, can very quickly result in ostracism and/or ongoing physical or psychological violence for the duration of one's sentence. (Halsey and Harris 2011: 75)

Conversely, the psychiatric nurse (prison setting no. 2) was more able to gain trust from the inhabitants. The male nurse was open to their stories and take account of their diversity. An older prisoner (in his late 60s), who exhibited signs of institutionalisation, would never participate in activities[5], except for the occasions when the nurse would cook a special treat for all of the inhabitants, and he would be one of the very few who assisted the nurse and genuinely enjoyed it. This prisoner contributed to this small community by doing some cooking (secondary generativity). For the most vulnerable individuals in prison, generative behaviour thus provided a stimulus in terms of their basic senses (sight, touch, hearing, taste, and smell) and is characterised by a low threshold. In the same way, language and music are examples of potential vectors for generative behaviour.

## Generative Behaviour Through (Grand-)Parenting

Based on his intra- and extra-mural experience, a middle-aged prisoner who dwelt on the impact of imprisonment on ageing prisoners' familial roles, touched upon (grand)parental generativity.

> R: [...] Come on, you cannot expect that your grandchild in prison... That's not feasible... That's an unrealistic expectation. With the result that certain roles who those people otherwise actually could fulfil... They sometimes have a photograph of their grandchild, but hey, that's it.
>
> I: Why is that impossible?

> R: Meh… I think that the reality actually is: if you as a parent get a visit from your children, you are very fortunate. And erm… Because you will always… Ehm, I mean, I expect that there the erm complicated situations… If the grandpa is in jail. Ehm, the son or daughter may still have some sort of loyalty towards the father, for one or another reason. Even if he or she has been the victim, that… that is sometimes complicated. But it might be there. But the in-law family, will also exercise pressure and then that grandch… […] Well, this hasn't received attention, but I am convinced that that is a factor, because I've seen in the psychiatric institution that you have the same phenomenon. That people still are in touch with their children, but with their grandchildren, that is very difficult, and they do miss that. Especially for older… Well, for an old person… the natural logical contact with someone of, whom is still really a child, are the grandchildren or greatgrandchildren, those things… There you have a little of that natural contact. Which of course is limited to… that you are in fact a sort of a 'glorified sugar uncle' (laughs). But it is a piece of life. (Interview Peter, March, 2016)

During their imprisonment, many older prisoners were completely cut off from any physical contact with their grandchildren or children, or their partner's (grand)children which were de facto 'adopted' and were considered as their own. Few participants still maintained significant contact with children and grandchildren. Those who were still allowed to maintain a form of contact with their grandchildren (either by parental or by judicial authority) actively withheld information about their whereabouts or status and voluntarily dismissed all likelihood of receiving (grand)parental visits. This non-disclosure was prompted in part by feelings of shame but to a much greater extent by protection of the (grand) child. When contact is established or maintained, it is limited to, in descending order of significance, telephone calls; written correspondence; and financial and/or material support. Imprisonment negatively impinges on the expression of (grand-)parental generativity.

With regard to one of Belgium's age-integrated prisons, a non-custodial staff member stated that she was once asked to organise a 'visit for grandchildren'. She continued that there was a lack of knowledge about the broader support for this proposal amongst incarcerated grandparents, and if so 'actually ought to be investigated': if there were a broader support, she suggested that they (the prison authorities) should consider if this 'grandchild visit' could be subsumed under the already organised 'child visit'.[6] This was argued not to be the case at the time of the study (Field notes P1, June, 2014).

Dorothy (65 years) stated that she was sure that a lot of men would want to see their grandchildren again. She explained why she believed that loneliness is much greater amongst male prisoners: 'I reckon that this applies even more to them: being strong and masculine, and that those things are thus not discussed' (Field notes P1, September, 2014). Despite how strongly she felt about the significance of maintaining contact, even Dorothy persistently concealed the true nature of her absence. Whilst elaborating on this topic, a number of prisoners mentioned that they felt as though 'prison is no place for children', that they are too young to comprehend or fully grasp the notion of imprisonment, and lastly that it might be a traumatic experience for them. Generativity might thus also be reinterpreted as the desire to protect children from prison influences—in some rare cases even themselves.

Within the domain of family ecology, grand generativity includes grandparenting and is characterised as a softer, indirect, and a less day-to-day connection to the life review (de St Aubin 2013: 244). As Chinen (2010: 82) writes: 'Many parents develop a deeper feeling of generativity vis-à-vis their grandchild, as they are less likely to apply the successes or shortcomings of the child to themselves' (own translation).

In his discourse, Peter touched upon a key issue within family ecology: that relationships become even more complicated when the incident, which led to the incarceration, occurred in an intra-familial context.

The data revealed how, for many prisoners, generativity was regarded as an intrinsic consequence of the prison experience. This is especially so for first time prisoner incarcerated in late(r) life, where parental and grandparental generativity, as well as their generative self-identity, was strongly jeopardised. For those at the beginning of their sentence (sentenced or on remand), the preponderance of the focus lay with the antecedent leading up to detention, rather than with the actual imprisonment, despite the latter being the result of the former. The following prisoner addressed feelings of shame as a consequence of his criminal act and as a consequence of his prisoner' status with a spoiled identity of character (e.g. Goffman 1963).

For Larry, one of the core participants serving a first-time shorter sentence, his identity was spoiled by being admitted to prison:

> I: What was important in your life? What was of value to you?
>
> R: My values? My household. My family. That is only logical, huh? Well, and the kids… Well, your family, huh.
>
> I: Your children?
>
> R: Children, and grandchildren, huh. That's everything. Yes. Your family… well, *the* family.
>
> I: Uh-uh.
>
> [silent pause]
>
> R: I'm ashamed…
>
> I: Why are you ashamed? Because you're here?
>
> R: Because I… Well… My family doesn't know. My children don't know. Only my friends, but they don't come visit me. That I did something stupid. Guilty of a dreadful piece of stupidity. Now I must pay for it.
>
> I: And why are you exactly ashamed of?
>
> R: I've said it once. I am insulted in my honour. Offended in my heart. I still have some kind of pride, huh.
>
> I: Yes. I'm just asking to know what's important to you.
>
> R: Ah yes. You live for this, to start a family. Ah yes. That is everyone's aspiration, right? (Interview Larry, October, 2014)

For Ivan (Interview, November 2014), generativity seemed consigned to the past (before and during detention). During the study, Ivan was encountered during a period of transition from generativity to stagnation. He had lost contact with his children and, very recently, with his wife, and he had given up his job with a trust function in prison. He could not deal with issues such as the aversion of his children, and by extension his grandchildren, and with the very difficult and complicated relationship he maintained with his wife, the victim of his crime. Whereas, at first, he still gave advice to fellow prisoners and prison staff based on his professional knowledge, focussed at gaining knowledge and worked in the prison, his ability to care for another person or to guide others slowly faded away. He was most concerned about his wife. It could be argued that the reported data supports the divergent body of literature that denotes generativity in terms of 'moments' and/or 'episodes' rather than 'stages' (Kotre 1996: 261-263; Kotre 1999: 23).

Ivan narrates a generative story imbued with joy and lasting gratitude catalysed by his encounter during his imprisonment with a familiar child, a distant relative who came to see him twice whilst he was in prison:

> R: [...] And that Ayda has a son named Mickey. And Mickey suffered from leukaemia. And uhm, for a whole year, I went every day to the University Hospital to support that little guy etcetera. He had, he had an intravenous drip for taking blood samples and so on, which needed to be done every day. But he was already crying when his patch needed to be replaced. And I have taught him to laugh with it and shaking blood tubes and so on. And he has visited me in here. On his own request, nota bene! I didn't want for him to visit me here. Uhm... he was perhaps five years old at that time. [...] 'Can Mickey come and visit?'. I've said: 'No. No children of uhm 5-6 years old'. And when he turned 9, I did say 'alright'. She had asked again. I said: 'he can come now'. So you mustn't wonder how... people, especially children, can be so grateful because you supported them through difficult times. Through difficult times... I went... I walked around with him, to god knows anywhere in the University Hospital, while carrying him around, with masks on of course, huh. So anyway, I thought that was nice.
>
> I: And why wouldn't you let him visit you previously?
>
> R: Too young. Too young. I find prison, um... not the... I find that it is something traumatising uhm for a child. Uhm. At 9 years old, they can comprehend a little more, so he understands that prisons... and so on. He also knows the reason why. So to say that children can appreciate something. And that brought me pleasure.
>
> I: How did you experience that visit?

R: Well, fantastic. That really uplifts your mood. He came with his grandmother. So, his mother, my sister-in-law, and my wife [the interviewee lights a cigarette].

I: And have you seen him since then?

R: I saw him twice. But now, uhm, my uhm… so Lena came with her sister, that is my wife. And Ida here. So, she came two times with the little one. My wife, for instance, she can't drive. (Interview Ivan, November 2014)

## The Prisoner Forum

The prison authorities of prison setting no. 2 commented, through the social service policy officer, that to ensure that older prisoners' views are heard, they planned to develop their prisoner forums to have an elected representative from each of the units, including from the wings holding older prisoners (cf. Moll 2013: 25). However, at the time of the data collection, there was no elected representative from the specialised wing. According to one of the prison governors, older prisoners on the specialised unit are given the opportunity to stand as candidate in the election, but there was little interest (Interview, Prison Governor, June 2016). The question arose whether the establishment of the existing prisoner fora is adequate for this purpose. The literature describes distinct older prisoners' forums (also known as 'senior forums') where the participation of older prisoners is often even stretched further:

> Forums may take other forms, such as prison-wide Measuring Quality of Prison Life (MQPL) meetings as well as wing-based general consultation meetings to which buddies and disability reps can be invited, as well as dedicated senior forums. (Age UK, 2011: 19)

Dorothy was the only prisoner who raised and addressed the issue of the prisoner forum[7] in prison setting no. 1. During her imprisonment, Dorothy was in fact elected to sit on the prisoner forum as a full representative but has pulled out after one year because she felt insignificant and unable to bring about anything for others: 'After a year, I told them [fellow prisoners]: I said "Listen people", I said "I am preparing that, afterwards I even make a report of it and everything, but you must understand that I'm talking against walls"…' (Interview Dorothy, October 2014). She not only encountered difficulties in addressing and conveying issues, but her proposals on behalf of younger and/or older prisoners (e.g. sports and activities for the over 50s such as yoga, canteen) were also disregarded. Dorothy's account reflected a growing distrust vis-à-vis the formal prison system that appeared to be hostile to the fulfilment of true generative desires and explains the rejection of the older population towards formal political activity and processes. Older prisoners felt less well represented and had the impression that their opinion was unvoiced or obsolete in the contemporary prison system. The generic prisoner fora were seldom commented upon the remaining respondents; it would be interesting for follow-up research to examine, to what extent older adult prisoners are informed of these bodies and/or included in them.

Similarly, acts of participation and representation (as acts of resistance) might also have the potential to develop a clear element of generativity, where activity is aimed at enhancing another's well-being. For instance, one older prisoner spoke about how he would tear prison underwear (with holes and worn-out elastic which he showed an example of in his cell) to pieces so that it could no longer be distributed amongst the inmates: he wore his own private underwear; thus his intention transcended his self-interest (Field notes P1, December 2014). In a similar way, Benedict (2009) terms the behaviour where inmates share food 'secondary generativity': 'Inmates do not have easy access to free society, so "giving back" to the prison community—other inmates—provides them with an opportunity to do good and to feel good about themselves' (Benedict 2009: 142).

## Generativity Through Relationship with Significant Other

Dale (male, 69 yrs) needed the money he earned in prison to support his younger sick wife, who lived in a foreign (developing) country and needed financial means for subsistence and treatment for disease. Deducting his necessities from his pay, he would send the remaining sum to her. To enable this, they invented these so-called 'kiss-calls': he would phone his significant other from prison every evening at a fixed time. They agreed that she would normally not answer these calls, unless there was an emergency or problem (Field notes P1, April, 2014; Field notes P1, November, 2014). He felt responsible for his wife and invented these calls to save money.

In his cell are many painted or drawn portraits of his own output of Alice, his foreign wife in early to middle adulthood. She had cancer and he would send money to her. He has ambitious plans (literally in the sense of construction plans, as well as figuratively speaking in the form of blueprints lying around in his cell) to build him and Alice a house in her country of origin (Field notes P1, November 2014).

Prompted by the age difference between his and his wife, Dale wanted to construct something to care for his younger wife and wanted to give something which he would be remembered by. Although he would speak very fondly of her as his 'life companion', he would restrict his verbal contact with her, in order to provide for her and ensure her medical treatment. They would however sometimes send letters. He received no visits from his family.

Some prisoners included in the study reported the absence of a significant other: they were either single, divorced, or widowed. For many, previous relationships had ended or were complicated as a result of the detention or its cause. It was noticeable that participants (M/F) who were still involved in a relationship often had partners who were chronologically younger than themselves. Several the male participants were involved in romantic relationships with a partner who was a lot younger (i.e. early to middle adulthood) in calendar years than themselves. These prisoners, still involved in a valid relationship, expressed generativity through their partnership or marriage.

## Creating a Product

Generativity was also expressed through means of creativity. A number of authors have already touched on various forms of generativity, which some suggest as acts of 'creation'. The prime example was given in Erikson's traditional contribution, where creativity and productivity were subsumed into generativity (Erikson 1950). Furthermore, McAdams has referred to Becker's creation of something 'heroic' ( McAdams 1988: 257–258), and lastly, Kotre, who discussed creativity under the cultural type of generativity, suggests that 'to be generative is likewise to be creative' (Kotre 1996: 12). He continues by commenting that creativity implies that something new is made and the act ends once the product is made. Generativity, however, describes passing along something old in which the self is infused and continuing to care for that product as it grows and develops (Kotre 1996: 11), thus 'all three points of distinction add up to greater duration in the generative act—more time antecedent to the moment of birth, more time following it' (Kotre 1996: 11).

Several participants expressed themselves through the arts, by playing musical instruments, writing, gaming, cooking, drawing, painting, sculpting, or crafting. Many participants took up drawing, painting, and writing books or stories as an intrinsic part of their imprisonment. They were successfully involved in creating a product; however, they were less accomplished in giving the product up as a key step in the generative process (McAdams 1988: 276). In most cases, their generative gift remained bound to prison walls and had at best benefited other prison actors. Conversely, a few participants succeeded to overcome the prison walls by benefiting the larger community. At the time of our fieldwork, Dorothy (65 yrs old) was in the process of writing a book on life in prison, which she referred to as 'hell'. At frequent intervals, she would send her daily personal notes to her publisher which would then be turned into chapters of her book.

Strikingly, many prisoners involved in this kind of activities were autodidacts, making them suitably qualified for technical generativity. There were instances when the prison setting appealed to prisoners' abilities: for instance, during the beginning of the research, printed postcards were received illustrated with drawings of prisoners by one of the prisoner-servants (65+ yrs); the signature and style of one of the participants could be easily recognised. The postcards also adorned the walls in a specific prison area. Sometimes, drawing contests were held amongst the prisoner population, and in one case, a talented older adult prisoner was invited to help with the decor of a theatre play, which he declined. Nonetheless, the way prison life is organised often militates against these forms of generativity, for example, the reduction of people by an officious librarian, the deprivation of resources (e.g. props), the prohibition on keeping musical instruments in a cell, and the abolition of courses that stimulate generative outlets. The examples that are given thus far mostly relate to the creation or production of something. However, generativity also comprised the renovation (Kotre 1996: 12-14) of existing products as shown by the following data sample where a prisoner casts a different light on old comic books of his time:

> [...] They have no librarian anymore over here.[8] The previous one eloped. [laughs] I do miss that... Because I was working on a study, well 'study'... Uhm, trying to compare comic strips from my time with those now. And I was working on that but, well yeah. (Interview Albert, February 2016)

This excerpt is a prime example of how older prisoners are at times de facto excluded from the general provision of services and programmes in prison.

## Belief System: Spirituality

Prisoners with religious belief systems believed in a similar yet different symbol of immortality as opposed to generativity, simply by living on in an afterlife. We identified religiousness, or spirituality, amongst the sample predominantly as an emotional coping strategy rather than a manifestation of generativity. However, that religion might involve, or include, generative elements (such as cultural generativity) is revealed by the interview with one of the older male prisoners. In the second of three interviews, Larry elucidates the role of religion in his life and by extension that of others:

> I: Are your children also religious?
>
> R: Well, of course they are also religious. They are baptised, they have had their confirmation, they have everything...

I: So that's important to you?
R: Well, that's my strength, ma'm. I can't help it.
R: You must have something, right? If you got nothing…
I: Yes, of course. Everybody has something.
R: You must hold on to something, huh? And those… And I find it… When you see [what's going on] with Islam. That's not catholic. That's enough, that fighting. You cannot convince me, Islam… […] Everyone [is entitled to] his…
I: Their own faith?
R: Each his opinion, hey. (Interview Larry, October 2014)

Larry saw it as paramount that his children were also religious or were at least brought up according to the spirit of Catholicism. In this sense, he wished to pass on traditions that were important to him.

## Generative Behaviour Through Life Story/Event—Narrative—Story Telling

From the first until the last encounter, Antonius repeatedly spoke of his biography, where diverse elements of generativity were rooted and expressed. In his primary personal encounter, Antonius spoke of several binders and stated that 'I have written down my whole life' (Field notes P2, January 2016). In his interview, he spoke of three binders, each one containing handwritten accounts that represented a specific event in his life and all on the theme of saving lives. He continued to speak about how he wanted to give a part of this autobiographical history (captured in one binder) to his son, which he gave to a requesting prison officer who promised to hand it to his son, but according to Antonius, he never did: 'And I got visitation and I said: "Well, I will give it, that book, that last part to my son". And there was a guard who said: "I will give it", and he never gave it' (Interview Antonius, March 2016). As a result, one of the four binders had gotten lost. Antonius further comments on the impact of this past event: 'So because of that, you don't have any trust anymore, huh' (Field notes P2, January 2016).

In almost every encounter with Antonius, he reiterated his narrative on saving the lives of four humans, of which one chapter had been lost by one of the prisoner officers. He starts to get cross: 'goddamn!' he shouts. He continues: 'And now, I will not start over anymore…' (Field notes P2, January 2016).

During his formal interview, Antonius once again raised the issue of the partial loss of his life story:

> I: Why won't you write it again? You don't feel like it?
> R: No. I do not care anymore.
> I: What are you going to do with the other (binders)?
> R: I leave them with me. When I get out, I will hand them to my son.
> I: Did he ask for them?
> R: I already gave a part, the first binder.
> I: It is interesting, huh, that you wrote that down?
> R: Almost an entire life story huh, apart from that which I gave to the prison officer. […] And he never gave that book. Pfft!! This is why, I have said, huh, that I don't do anything here, nothing anymore, I don't care anymore, do you understand me? Pfft!
> I: But at least you have the other ones?
> R: Yes, I take good care of them. It says that I have rescued three people in those books that I have written.
> I: That is something you must never forget.
> I: But I will definitely not give them anymore to the guard. Goddamn! Yes, yes. Gosh, dammit! [brief pause] You take..? I have written to that governor; I have written four years and now I lost it. Fool. And that calls himself go-ver-nor [disparaging tone] … Pfft! (Interview Antonius, March, 2016)

It is worth noting that Antonius began writing his life story at the beginning of his incarceration, which was in a different prison (ironically, prison setting no. 1 in this research) from where he was transferred. In his answer to the question of how he experienced the former prison, Antonius refers to malpractices on the part of the prison governor: although he wrote his complaint multiple times to the prison board, he never received an answer, 'yet we are all humans' he lamented. He continued: 'Just because you are in jail, doesn't mean you shouldn't receive care' (Field notes P2, January 2016). 'And I have already written four or five letters to the prison governor. And I never received no answer. Well, I say: "all of you can kiss my ass", in my mind, huh. Not too loud... You understand?' (Interview Antonius, March 2016).

When examining the content of his written and narrated life story under closer scrutiny, the theme of 'rescue' was paramount, and the act of saving or rescuing the life of others runs as a continuous thread through his different stories:

I: Do you still have a recollection of the story in your mind?

R: Yes, yes. But it is in my head hé. It was in [a small town in the North of Belgium]. The fourth man that I, he was hit with electricity. And I called out to one guy: "Pull out the plug!", and he ran away. I jumped off the scaffold, the platform, I pulled the plug. That electricity, the machine was still running, 6000 revolutions per minute [...] He dropped the machine and I was able to catch him and lay him on the floor. The doctor said, we drove over there, and the doctor said: 'if you would have left him another minute, he would have been dead'. But I was working there illegally. It had to be kept silent. Both the bad and the good, huh. And the people who came, they thought that it was another man who saved him, and I never spoke of it. I never said: 'It was me'. No no."

I: But you yourself do know.

R: Yeah. I saved four people.

I: Do you sometimes look back at that?

R: Yes. One day, I came from work, illegal work, chores, huh. And my neighbour went like: 'Antonius, they phoned, you must immediately go to the hospital'. So I went to the hospital, but I didn't know why at that time, huh. And there I was called separately, and they collected a blood sample from me, and that was for a new-born child.

I: Because you had such a rare blood type?

R: Yes. [...] And twelve years later, I visited a nephew of mine in the hospital, and a woman entered. [...] And she said: 'I am the mother of that child whom you've donated blood to twelve years ago.' I said: 'I will come and visit your son once'. It was his solemn Communion, and I gave him 500 Belgian Francs.

I: Well, that is quite the story.

R: [..] Well, yes. I was proud of it. And the first one who I saved; he fell into the water. Little Alex was his name. I told him to leave everything alone and then he tried to fish with those poles and he fell into the water. I immediately jumped after him and I got him out.

R: And the remaining person? You said there were four?

I: That child....

I: And a friend. He worked in the coalmines of Houthalen, do you know that?

I: Yes, in Limburg? [i.e. a Flemish-speaking county in Belgium]

I: Yes. And we worked there. And he got a letter from his girlfriend that she ended the relationship. And I received a letter from her, that same day... And out of anger, he climbed on something to receive an electrical shock. I said: 'Curtis, get down. You are a coward'. He was no coward, hey! That was a word to get him down. 'I will slap you in the face', I said. And he, he came down already beating the air, and I went 'BAM!', on his nose, two black eyes and a broken nose. It was on a Friday, because I remember we went home on Saturday. And in the morning I said, 'Hey Curtis, get up', 'mope mopemope', he didn't want to. And I left, I thought: 'he will manage'. [...] And when he got off the train, I saw him with two black eyes. 'You must say in our town that I got a rock on my helmet and a bleeding nose and black eye'. I said 'Yes, sure'. And it was left at that. (Interview Antonius, March 2016)

Antonius' behaviour is a prime example of the combined modes of generativity. Illustrated by the written and spoken accounts of his life story, Antonius acts as the hero or saviour, or as an *agentic* dispositional of generativity in the narrative(s) he constructed. However, he was strongly committed to handing over these heroic narrative stories to his son, thus literally 'letting go of his creation' after his own image by donating it in the *communal* dispositional of generativity. By giving a binder (part of his story) to his son, he aimed to provide a legacy to remain within the realms of the family. That, precisely, is where the fulfilment of the generative process is compromised by the prison institution. Losing one's generative product might therefore suggest a lack of respect, either deliberately or otherwise, for prisoners' personal possessions, and/or the complete disregard of the symbolic meaning and value of them. The cited excerpts of the research material reveal how generativity could be generated during, or even in response to, imprisonment and/or ageing, by the creation of something heroic, but could not be completed, due to the actions or negligence of actors within the prison structure.

During the last interview, whilst recalling one of his best memories, Antonius recounted fishing trips during which he taught his children and grandchildren how to fish. He shared his experience of taking them, on trips out to sea on his Zodiac boat. These memories of generative activities constitute his response at the same time to a query about what he found the worst thing with being imprisoned (Interview Antonius, March, 2016).

## Generativity Towards (in the Vicinity of) Objects

In their study of infertile married men, Snarey et al. (1987) identified a 'parentlike devotion' to nonhuman objects such as a pet, car, garden or house, as a substitution for those who expressed an intermediate interest in generativity. In the same way, recent studies suggest that generativity can, and does, exist, even amongst people with advanced stages of dementia (Doyle et al. 2015). Following this, instances of cognitively impaired or otherwise older prisoners selecting an object which they could focus their generative care upon were conspicuous in the findings.

Anna, for example, a woman aged 73 and diagnosed with Alzheimer's (and who passed away during the fieldwork), was 'interned' in prison as a mentally ill offender. Whilst strongly enjoying my company, sitting on the bed next to her, Anna talked about how she cared for her cuddly doll 'Poppemieke' sitting on the desk in her cell. She was clear that it was a girl, and that it was a doll, of which she had been in possession of for two or three years. She would remove, but not sleep with, the doll, because she was afraid it might get broken (Field notes P1, August 2014). Although this doll[9] strongly symbolised her need for (meaningful) contact, (Anna was largely cut-off from her sister (66 yrs) due to the distance from the penitentiary), it also represented an endeavour to gain a—albeit minor—purpose by looking after and caring for something. However, for most prison staff, this behaviour was tolerated but represented as something 'negative' and 'erratic', strongly associated with the deterioration of a person who had reverted to childlike behaviour, rather than as a meaningful developmental task.

## Generativity upon Release, Prompted by Imprisonment

Experiencing imprisonment can have a significant effect on older prisoners' generative aspirations and expectations. Strikingly, for some of the participants, the harsh experience of imprisonment prompted future aspirations of generativity upon release. In this respect, generativity manifested itself by seeking redemption and/or revealing generative concern for others in the face of vulnerability.

In line with the findings of Benedict (2009), some participants expressed strong desires (to continue) to be generative upon their release. Larry, a timid 72-year-old first-time prisoner, was admitted to prison late in life and was clearly intimidated and overwhelmed by his incarceration. His choice of words revealed the order of intent in his underlying thinking—in which things needed to be done. He talked first about making a pilgrimage to one of the most popular pilgrimage sites, which he had to do before helping older people. Guided by extraordinarily strong feelings of solidarity, Larry expressed hope and a fervent desire to help other older nursing home residents, whom he viewed as equal to himself, on the basis that they were all 'being locked up'.

During the second interview with Larry, he touched upon his generative aspirations for the future:

I: During our last conversation, I asked you what you still would like to do with your life, to which you answered: 'helping people'?

R: Well, first I'll go to Lourdes. To Lourdes [i.e. a well-visited pilgrimage site]. And then with the elderly. Helping [them]. Going to places… Get out of the house, huh. Inviting them to… go over there. Drive… To the basilica… Just short distances. I'm gone and she's [the person being visited] gone… That is what I would like to do.

I: Have you done that before?

R: That is what I would like to do… So, people with…

I: How did you come to this idea?

R: Just… Here… In here it makes me feel… in lieu of other people. They are also locked up. Hey, and if I make the comparison… [I want] That these people are outside the house for once. For that reason. Well, I do think so.

I: That's nice.

R: I've experienced that myself, indoor time. And that is what I'd like to do, yes… People…

I: People who haven't got anyone?

R: Exactly! I have a car. And I can drive.

[…] Many people who are stuck inside… that they come outside for once. And making trips. […] In a nursing home… Where there are a lot of people, huh? And they don't even have to pay me. No? Voluntarily. I don't need anything. Not even… my car. I drive with it, huh. Thus, I also derive pleasure from it. And hey, to take two people is no big deal. I do not care. I don't care about money.

I: You do not worry about money?

R: No. No. No.

I: Is there anything you do worry about?

R: When I… When I die, I won't take anything with me, huh. (Interview Larry, October 2014)

In a similar way, Dorothy's account reflects her combination of writing and illustrating children's books for publication, as well as providing financial generativity by making donations (Interview Dorothy, October

2014). Dorothy had taken her case to a supra-national judicial body, claiming that her right to a fair trial was violated—at the time of writing her complaint was being taken under consideration. She passionately believed they would find in her favour and thereby impact upon her goals for the future. It is striking that in Dorothy's account, there is an element of selectivity present vis-à-vis drug users and child neglect:

> I: How would you like to spend the rest of your time?
>
> R: Just continuing like I'm doing now. I reckon. And having more activities, uhm… Must it ever been that I have to stay longer or… Another year, and then I've done my third. Then… erm… I would definitely start writing books for children. Indeed.
>
> R: So you have designed a way forward. How do you see yourself in the future?
>
> I: My future? That's spoiling my husband! Hehe. [laughs] And I reckon, again, thus… Uhm… If there is the possibility, thus… I always say 'with the money of the Belgian State, wait a minute. Then I would definitely, definitely go traveling. Indeed. And I'm going to make sure that I erm can help children, but in Belgium itself. Yes. So, but being able to provide it myself. For example, in poor neighbourhoods, [for] Saint Nicholas[10] or anything else. So I've always promised, that, a third of the amount that I will receive, I will use for… erm… the less fortunate, let's say…
>
> I: And why do you choose children?
>
> R: Because, first of all, they are so innocent. And secondly… My, my thingamajig goes out to children, or to very old people. Thus, you'll know thus they have… Ghosh… I still remember in the old days we'd do that at school. Uhuh. Thus, bringing coal to older people who couldn't afford their coal. Huh, all such things. Therefore, those are two…

> I think that the rest, well, they must make sure that they can work. Do you want to work, then you can work. It is as simple as that. In my entire lifetime I have done nothing else [scratches her throat] … At that time, when you still changed your job for 50 cents. I can understand that it's not like that anymore. But if you want to work, you can always work. Very simple. Thus… that group… If they don't do it, and they are in misery; in certain

cases, it is simply because they are too lazy to get out of their bed. But like small children and so on. Furthermore, you must not forget that if you can give something to small children, the people they will also feel a lot better. And, well, if I then, for example, pick out some families to help them with the clothing of their children. Even if it is… Even if it is with their shoes alone, for example. Huh, just think how much it costs to get winter boots. And this and that. Well, if I can, take care of that, then the family will also be happier. The mother will always feel happier herself, huh, in normal circumstances. I am anti-drugs. Thus, er… I'm going to examine that first. Hehe. And uhm… that is what I would like to do. I've always sworn that. A third of the amount of money that I'm going to receive—if it's likely to stay like this, on the contrary, I've heard that it will even add up—is a nice sum. And I no longer need a house and so on, because then the children would fight and I don't agree with that, so…. (Interview Dorothy, October 2014)

## Life History Accounts

### Generativity Through Work and Productivity

For Westermeyer (2004: 36), generativity must have 'a clear responsibility for others, successful in work, and achieving intimacy'. In his interview, Frans (a male prisoner aged 68 and diagnosed with a decline in cognitive function) clearly described the way driving with his truck, his former job, made him feel:

R: […] Work… Work a lot. I miss it… My truck, I miss it.

I: How did that make you feel?

R: I drive with the truck and you are response… [looking for the right word in Dutch].

I: Responsible?

R: Yes. (Interview Frans, March 2016)

For Frans, nothing could give him the same feeling as driving his truck; it was the best experience in his entire life as well as the thing he missed most in prison, and it embodied his future aspiration(s). Although Frans was diagnosed with dementia, shuffled up and down the hallway, and had problems with his joints and knee, he repeatedly, and unsuccessfully, asked the prison governor to serve or work in prison, even to drive around the courtyard with his truck. The refusal of his requests made him feel as if he could no longer contribute. His accounts are consequently interspersed with stories of his previous work, symbolised by his truck.

Most prisoners, regardless of whether they did work in prison or not, made requests to 'work' or to contribute somehow in prison. To some extent, it is arguable that the large representation of prison servants (fatiks) amongst the older adult prisoner population might be related this phenomenon.

## The 'Dark Side' of Generativity

Whereas Erikson and others invariably described generativity as an ideal (Schoklitsch and Baumann 2012: 264; Kotre 1996), a few scholars have pointed to what they term as the 'dark side of generativity' (Kotre 1996: 9). Kotre (1996) argued that generativity's 'multi-fold capacity for perversion has been overlooked'. Instead of leaving a veritable legacy with virtuous intent, one might bequeath a heritage of active destruction, propagating evil. He therefore advances the view that generativity ought to be seen as 'an impulse that can be channelled into vice as well as into virtue' (Kotre 1996: 9).

Previous research on intergenerational incarceration has focussed on the healing power of generativity in the relationship between formerly incarcerated fathers and their (incarcerated) sons and in particular how fathers assist their sons through and beyond prison (Halsey and Deegan 2012). In this sample, only one such father over the age of 65 was willing to participate. He spoke very little about his son, who had been released but had already been recalled to (another) prison. In the past, they resided at the same unit, after being convicted for the same crime. The goal of this prisoner ('doing business') and the legacy he wanted to leave might raise questions about the toxic generative legacy he might be passing on. Generativity in this instance could be regarded as the bequeathing of intergenerational damage by portraying deviance and offending behaviour as a desirable, or optimal, life choice.

Another example could also be found in one of the research participants with a predilection for mythomania who had a past and a present of disseminating extreme and hazardous thoughts on alternative therapy to other human beings, especially followers. He was strongly convinced that he was spreading 'good' and that he was the personified perfection, on a crusade battling against traditional views and ignorance. The same observation might hold true for some paedophiles (or others), who argue that they are carrying out beneficial actions that are criminalised by the government.

Whilst the very small body of research that exists on old-age generativity in custodial environments does not correlate generativity with subsequent offending (cf. Maruna 2001), it does however seem to suggest connections between generativity and peaceable behaviour in prison, as generativity-minded older inmates seek helping rather than harming opportunities behind bars (see, e.g. Benedict 2009: 152).

## Growth and Development of the Older Prisoner

This chapter shows that older prisoners still have developmental tasks and roles to fulfil, which is commonly unacknowledged, unrecognised, and certainly unaddressed. It is unsurprising, given the lack of attention to older prisoners more generally, that little or no reference has been made to older adults' need to fulfil their generativity needs. The experiences of the prisoners in this study are imbued with both agency and communion, albeit in varying composition. The unilateral view of older prisoners as passive receivers and incomplete or meaningless adults is not reflected in the evidence. On the contrary, they emerged as competent social actors who can shape their lives and who seek to be involved with, and contribute to, the development of others and future generations.

In addition, the data strongly suggests that opportunities for acts of generativity are significantly contoured by the prison context. The findings show that whilst prison might very occasionally make a positive contribution to generative behaviour, its impact is overwhelmingly negative, either by discarding, challenging, or even rescinding opportunities for generative activity. By its very own nature, the constraints of imprisonment limit opportunities for prisoners to give anything back to the outside society or community and/or to express secondary generativity within prison. There are many influencing factors such as the prison regime (e.g. segregation) and prison practices and attitudes from staff. Through such structural features, prison exposes elements which may be regarded as the dark side of generativity. Or, put simply, prisoners are seen as unable to contribute something desirable to the community, a culture internalised by some individual prisoners, potentially resulting in stagnation.

The implications of this suggest that it would be beneficial to look for meaningful use of time such as caring for animals (Halsey and Harris 2011), or the better use of 'old heads' (Maruna et al. 2004: 146). Supporting prisoners to accomplish developmental tasks, rather than purposeless activation of older prisoners, must be the clear and optimal policy trajectory.

## Notes

1. Jung divided life into the first and the second half, with four parts of the life course (i.e. childhood and youth for the first half of life; and midlife and old age for the second half of life). He advanced that the second half of life has the specific developmental tasks, including the process of individuation: the process of self-realisation ('becoming oneself') through which an individual integrates conscious and unconscious parts of the self (Kelleher 1992).
2. Kotre's original work was published in 1984. In the edition of 1996 (xvii), he wrote that he changed but one thing: he added a paragraph on the combination of agency and communion.

3. It must be noted that Joan Erikson continued the work of her husband (Erik Erikson) after his death and added this ninth stage of development when she was experiencing very old age herself (from a chronological perspective; she was post 80s to early 90s) (cf. references to her own experience with the final stages of life in the introduction of the extended version of *The Life Cycle Completed*).
4. This name refers to a senior structured living programme housed within a Nevada prison. See for more information Harrison (2006).
5. This research is no exception. The person did not want to participate in an interview and was very closed off. It even took a while before he would hold out his hand to be shaken. He indicated that it was nothing personal and he was not against the presence of a researcher but that he would never do interviews.
6. Regulated by Ministerial circular letter no. 1715 of 5 July 2000 on the protection of the affective relationships of prisoners with their environment, unpublished. The Belgian prisons must organise each month at least one activity for parents who are in prison and their children. As it remains unclear as to how grandchildren resort under these guidelines, children can visit their grandparents during normal visits. In the segregated prison, the visit is organised in the hallway.
7. Article 7 of the Basic Law of 12 January 2005 on Prisons and the Internal Legal Position of Prisoner (2005). This body is also denoted as 'gedetineerdenoverleg' in Dutch.
8. There was only one librarian for the general library, which was inaccessible to most prisoners in the special unit of prison setting no. 2 due to inadequate infrastructure (e.g. stairways). There was however a small library (a cabinet in the hallway on the ground floor), which offered a smaller range of books. During the research (and at that time already ongoing for well over six months), library services were not available for older prisoners, since they no longer had a 'librarian' (previously done by an inmate) who would occupy himself with the administrative and logistical part the loan and return of items.
9. From a Latourian perspective, it could be argued that this doll is an 'actant', as it replaces a real human body and makes others (actors) act, and it is regarded as a 'toy-partner for communication' (Vakhshtayn 2014: 419). Bruno Latour himself has apparently borrowed the concept of 'actant' from A. to G. Greimas semiotics; it illustrates an object's ability to act independently within the social world (Vakhshtayn 2014: 419).

10. This pertains to a tradition in the low countries. Saint Nicholas (in Dutch: 'Sinterklaas') is the main figure of the eponymous annual folk festival (traditionally for children), which is celebrated on the fifth and sixth of December in Belgium. To put it in a nutshell, the story prescribes that children who have been good all year will receive presents from the Saint, whereas children who have been bad all year are given coal, or even worse are lashed with a rod.

## References

Bakan, D. (1966). *The Duality of Human Existence: An Essay on Psychology and Religion*. Rand Mcnally.

Basic Law of 12 January 2005 on Prisons and the Internal Legal Position of Prisoners, Official Bulletin (2005).

Benedict, R. (2009). Giving Back Not Giving Up: Generativity Among Older Female Inmates (Doctoral Dissertation). University of Tennessee. Retrieved from http://trace.tennessee.edu/utk_graddiss/562.

Chinen, A. B. (2010). *Nog lang en gelukkig / druk 1*. Houten: BohnStafleu van Loghum.

de St Aubin, E. (2013). Generativity and the Meaning of Life. In J. A. Hicks & C. Routledge (Red.), *The Experience of Meaning in Life* (pp. 241–255). Springer Netherlands. https://doi.org/10.1007/978-94-007-6527-6_19.

Doyle, P. J., Rubinstein, R. L., & de Medeiros, K. (2015). Generative Acts of People with Dementia in a Long-term Care Setting. *Dementia (London, England), 14*(4), 409–417. https://doi.org/10.1177/1471301213498246.

Erikson, E. H. (1950). *Childhood and Society*. W.W. Norton & Company.

Erikson, E. H. (1993). *Gandhi's Truth: On the Origins of Militant Nonviolence*. W. W. Norton & Company.

Erikson, E. H., & Erikson, J. M. (1997). *The Life Cycle Completed*. W.W. Norton & Company.

Erikson, E. H., & Erikson, J. M. (1998). *The Life Cycle Completed (Extended Version)*. W. W. Norton & Company.

Goffman, E. (1963). *Stigma: Notes on the Management of Spoiled Identity*. Penguin Books.

Halsey, M., & Deegan, S. (2012). Father and Son: Two Generations Through Prison. *Punishment & Society, 14*(3), 338–367. https://doi.org/10.1177/1462474512442315.

Halsey, M., & Harris, V. (2011). Prisoner Futures: Sensing the Signs of Generativity. *Australian & New Zealand Journal of Criminology, 44*(1), 74–93. https://doi.org/10.1177/0004865810393100.

Harrison, M. T. (2006). True Grit: An Innovative Program for Elderly Inmates. *Corrections Today, 68*(7), 46–49.

Kelleher, K. (1992). The Afternoon of Life: Jung's View of the Tasks of the Second Half of Life. *Perspectives in Psychiatric Care, 28*(2), 25–28.

Kopera-Fry, K., Harrison, M., Iribarne, J., Dampsey, L., Adams, M., Grabreck, T., McMullen, T., Peak, K., & Harrison, W. (2012). Reflections on Being an Older Adult in Prison. *Aging and Society: An Interdisciplinary Journal, 1*(4), 1.

Kotre, J. N. (1984). *Outliving the Self: Generativity and the Interpretation of Lives*. Baltimore: Johns Hopkins University Press.

Kotre, J. N. (1996). *Outliving the Self. How We Live on in Future Generations* (2nd Edition). Norton Paperback.

Kotre, J. N. (1999). *Make It Count: How to Generate a Legacy That Gives Meaning to Your Life*. New York: Simon and Schuster.

Levinson, D. J. (1978). *The Seasons of a Man's Life*. New York: Knopf.

Maruna, S. (2001). *Making Good: How Ex-convicts Reform and Rebuild Their Lives* (Vol. XIX). Washington, DC: American Psychological Association.

Maruna, S., LeBel, T. P., & Lanier, C. S. (2004). Generativity Behind Bars: Some 'Redemptive Truth' About Prison Society. In E. D. S. Aubin, D. McAdams, & T. Kim (Eds.), *The Generative Society: Caring for Future Generations*. Washington, DC: American Psychological Association.

Maxfield, M., Greenberg, J., Pyszczynski, T., Weise, D. R., Kosloff, S., Soenke, M., Abeyta, A. A., & Blatter, J. (2014). Increases in Generative Concern Among Older Adults Following Reminders of Mortality. *International Journal of Aging & Human Development, 79*(1), 1–21.

McAdams, D. P. (1988). *Power, Intimacy, and the Life Story: Personological Inquiries Into Identity*. New York: Guilford Press.

McAdams, D. P., & Logan, R. L. (2004). What Is Generativity? In E. D. S. Aubin, D. P. McAdams, & T. C. Kim (Eds.), *The Generative Society*. Washington, DC: American Psychological Association Press.

McAdams, Dan P. (2001). *Coding Autobiographical Episodes for Themes of Agency and Communion*. https://www.sesp.northwestern.edu/docs/Agency_Communion01.pdf.

McAdams, D., & de St Aubin, E. (1992). A theory of generativity and its assessment through self-report, behavioral acts, and narrative themes in autobiography. *Journal of Personality and Social Psychology, 62*(2), 1003–1015.

Miranda-Chan, T., & Nakamura, J. (2015). A Generativity Track to Life Meaning in Retirement: Ego-Integrity Returns on Past Academic Mentoring Investments. *Work, Aging and Retirement, 1*, 1–14. https://doi.org/10.1093/workar/wav025.

Moll, A. (2013). *Losing Track of Time: Dementia and the Ageing Prison Population: Treatment Challenges and Examples of Good Practice.* https://www.mental-health.org.uk/publications/losing-track-time.

Schoklitsch, A., & Baumann, U. (2012). Generativity and Aging: A Promising Future Research Topic? *Journal of Aging Studies, 26*(3), 262–272. https://doi.org/10.1016/j.jaging.2012.01.002.

Snarey, J., Son, L., Kuehne, V. S., Hauser, S., & Vaillant, G. (1987). The Role of Parenting in Men's Psychosocial Development: A Longitudinal Study of Early Adulthood Infertility and Midlife Generativity. *Developmental Psychology, 23*(4), 593–603. https://doi.org/10.1037/0012-1649.23.4.593.

Vakhshtayn, V. (2014). On Microsociology of Toys: Material Turn and Non-Symbolic Interactionism. *International Journal of Social Science and Humanity, 4*(6), 416–420.

Westermeyer, J. F. (2004). Predictors and Characteristics of Erikson's Life Cycle Model Among Men: A 32-year Longitudinal Study. *International Journal of Aging & Human Development, 58*(1), 29–48. https://doi.org/10.2190/3VRW-6YP5-PX9T-H0UH.

# 7

# Conclusion

## Knowledge on the 'Older Prisoner'

Throughout this book, we have viewed growing older not only as a mere pathological process but also as an existential one (Baars 2012). The former view is predominant in a culture that cultivates the malleability of the body and mind and the manageability of the ageing process through, for example, medicalisation. This presupposes that ageing can be intervened in or otherwise (re)formulated: it renders 'ageing…[as]… something that is to be cured or treated'. This premise is strongly reflected in the prevailing discourse on the older prisoner. More specifically, it is implemented through various health screening initiatives and proposals, inter alia: the concept of 'accelerated ageing' among prisoners and 'activities of daily life' measurements and, for example, 'triage-tools' (Goulding 2013: 30).

Assumptions about what it means to 'age' have strongly informed the research upon which this book is based. Two underlying views can be distinguished in thinking about old age (De Motte 2015: 33f.; Fenton and Draper 2014: 3; Godfrey et al. 2004: 2). The first view builds on the 'deficit model' where ageing is perceived to be a process of decline, loss,



D. Humblet, *The Older Prisoner*, Palgrave Studies in Prisons and Penology,
https://doi.org/10.1007/978-3-030-60120-1_7

and social withdrawal. The second view can be summarised as the 'heroic model', whereby old age is framed as being active, fit, healthy, and taking place in a phase of life full of opportunities that young(er) persons are deprived of because of their responsibilities. This dual view on ageing feeds into the idea that 'life has already been lived' at older age and that later life stages are merely a prelude to death, which can only be overcome by a pursuit of youth-oriented goals. This has resulted in a bifurcation, reflected in the prison praxis—a gerontology of the usual, a '*normal* and *pathological* ageing' (Rowe and Kahn, 1987, in Ferraro 2013: 9). Thus, prisoners with signs of 'normal' ageing are to be integrated in the general prison environment, whereas prisoners whose self-management may be less obvious are labelled as 'pathologically' ageing and segregated from other inmates. In this respect, prison is a *heterotopia* which reflects societal norms (Goetting 1985; Foucault 1984). An alternative scientific discourse, however, questions the distinction between normal ageing and pathological ageing (Derkx 2011: 162). It rejects the heterogeneity and the modifiability of 'normal' ageing and considers its implications for harm and risk (Ferraro 2013: 9). The older prisoner is thus allocated to one or the other prison environment based on criteria that are underpinned by a questionable duality. This way of thinking about, and handling, the older prisoner and ageing is characterised by a high degree of intra-individual unilaterality and inter-individual duality, thereby reinforcing an ageist paradigm.

In line with this, this book has revealed how a relatively 'novel' analytical category of 'the older prisoner' has been constructed, whereby the older prisoner is identified as 'the other'. The 'older prisoner' has not seldom been defined in terms of disability, illness, and/or infirmity, whilst our research shows that the subjective and objective assessments of ageing are very complex and subject to great variability and change. As a result of this, the older prison population is strongly heterogeneous. This heterogeneity has been predominantly established in terms of *incarceration history* by others, resulting in a particular typology of older prisoners: recidivists with prior prison experience, those who have aged as a result of very lengthy sentences, and lastly persons admitted late in life with no prior incarceration history. However, in describing an older adult prisoner population from a holistic perspective, it is important to acknowledge the variability in pathways to both *prison* and *ageing*. Moreover, the underlying mechanisms suggest that prisons are increasingly pursuing the route of categorisation and truncated differentiation, whereby the general prisoner population is increasingly segmented into smaller categories. Gerontological insights have brought to light the 'intra-individual variabilities of age' that are left unaddressed in the penal field. Building on Wahidin's contention that the absence of attention to 'elders' in the criminal justice system could be regarded as a latent form of ageism (see, e.g. Wahidin 2004: 10), the research suggests that the question of *whether attention is being paid to older adults in the prison system* gradually should be replaced by the question, *in which way are these individuals receiving attention*?

We know now that ageing must be considered as a complex interplay of different processes, wherein one's chronological age constitutes merely one of many elements (Baars 2012). The study has shown that chronological age is *in itself* of little significance within the actual prison context, whilst other various other dimensions of age have been identified as more relevant. Although there is a strong interaction between most dimensions of (old) age and the prison environment, some are more shaped by the individual (such as the biological age, accelerated ageing, pain quotient) than others that involve a more prominent interaction with the prison infrastructure and regime (through functional and social ages). In scrutinising the different age dimensions for the core participants in the two prison settings, we found that older prisoners with similar age profiles were present in both the integrated and the segregated prison settings.

By drawing on self-assessments, the examination of the biological age of these prisoners showed how functional age did not necessarily correspond with biological age, indicating that prisoners with biological impairments could function well within a *certain* prison context and vice versa, although there was a surprising overrepresentation of individuals with functional limitations in the second research setting, where social age was constructed in terms of one's familial life events, engagement in prison work, and social habits. Most prisoners appeared to be in line with their social clock.

The sample differed greatly in the degree to which their psychological age has increased with chronological age. Having reached a high(er) cognitive age was most evident in the case of prisoners suffering from cognitive impairments. In both research settings, prisoners identified as those with cognitive impairments were encountered, and although chronologically younger prisoners were not exempt, those advanced in age were more liable to suffer with cognitive decline associated with ageing. However, most older prisoners were found to rely on higher emotional ages which tended to elicit feeling younger than their chronological age denoted.

## Impact of the Prison Environment on Older Adults

The emergence of the 'novel' category of prisoner has been almost simultaneously accompanied by the construction of a wider scope of the 'pains of imprisonment'. This book has revealed that the way these pains are experienced is influenced by the various 'dimensions of age' of prisoners. Thus, the following phenomena are argued to be of significance in the experience of prison whilst growing older: meaningful relationships with the self and others; a body and mind that is liable to change; approaching the end of life—i.e. all external or internal events or signals that draw attention to or are reminiscent of; and the growing vulnerability that is accompanied by growing older. All of these make meaningful additions to the traditional 'pains of imprisonment'. Similar to Goffman (1961) who described the impact of total institutions on its inhabitants as role dispossession, de-individuation, and mortification of self, this research shows that imprisonment creates a significant threat to meaning for individuals in later life, as ageing becomes accompanied by a loss of roles. With a more limited window of opportunity that exists for older adults to maintain social roles, it is thus argued that older adult inmates find it even more difficult to attain meaningful roles, either by old age or by captivity per se. The findings thus illustrate the lack of clearly defined roles, norms, and opportunities for older adults that could replace lost roles and create a new sense of purpose in life (Pinquart, 2002: 91). This holds even more true for prisons, as these environments have never been designed to house an older population.

## Material Prison Environment

Elements of *thoughtlessness* and *thoughtfulness* regarding the material prison environment were established in both settings. In Chap. 3, we have argued that the material environment shaped for older prisoners in the age-integrated first prison setting can be regarded as the epitome of institutional thoughtlessness. The remoteness of the building, the withdrawal to the constrained prison cell, and the difficult access to other spaces of the prison institution amounted to an amplified loss of liberty and autonomy for some older prisoners (Sykes 1958: 65ff.). The typical functional profile of older prisoners housed in the first setting accounts for the finding that an overwhelming majority of its inhabitants were shown to be able to make their way around the prison environment unaided, walk reasonable distances, and manage the stairs fairly. However, the material environment was unsuitable for prisoners who were, despite their chronological ages, were functionally advanced in age. This prison setting provided an excessively restrictive and non-supportive environment for prisoners, thus limiting freedom of mobility and physical independence, in ways most likely to adversely affect captives with high(er) functional age. It is certainly arguable that those with lower functional ages might be at (greater) risk of becoming functionally impaired as a result of the environmental conditions. In the same vein, the lack of personal ownership and of physical comfort in the older prisoners' material prison world emerged as more generic alternate pains of imprisonment within this setting.

The second prison setting, a gero-medical prison unit underpinned by a segregated approach, hinted at thoughtfulness. Whilst the physical independence prisoners housed here was hampered by their high(er) functional ages, it was compensated for in terms of a great(er) freedom of mobility and a less restrictive physical environment. In contrast to the first setting, the material physical environment was more oriented to support and compensate for the declining competencies of its ageing residents. In terms of thoughtlessness, the overall prison architecture was poor in meeting the needs of prisoners with high(er) functional dimensions of age. Whilst these prisoners were less confined *within* the unit, they were more confined *to* it and with a distinct the lack of privacy. Across both research settings then, the study exposed a strong interconnectedness between the functional age of (older) prisoners and the material prison environment.

The findings have further highlighted some of the 'material constraints' of the prison environment shown by Mann (2008; 2012; Mann 2016 which refer to 'limitations on choice and action imposed by the physical capabilities of the human body, plus relevant features of the physical environment' (Giddens, 1984: 174-6 in Appelrouth and Edles 2008: 761). The empirical data in the findings suggested that older prisoners were often a less empowered group of prisoners, especially with regard to those who had spent a long(er) time behind bars and were cognitively impaired. Hence, one of the strongest recommendations to emerge from this research is that the prison as an institution should show greater sensitivity towards the impact of existing structures on those with high(er) psychological ages.

## Social Prison Environment

The fourth chapter extensively detailed the vertical and horizontal relationships that shape the social prison world for older prisoners. It was striking that both older prisoners and prison staff referred to 'the system' when speaking of the prison, rather than in relational aspects vis-à-vis individual persons. The internal logic of the prison, reflected in the prison staff culture and the overall regime, was found to support the notion of *institutional thoughtlessness* in the adherence to the 'principle of sameness' and the way older prisoners were typically expected to conform to this principle. In addition, the findings revealed that older prisoners were sometimes subject to 'othering', in the form of negative stereotyping of characteristics intrinsic to old age. Thus, they become subject to a 'double bind' of sameness and otherness, which is indicative of ageism in prison that cuts across different dimensions of age.

Older prisoners were often treated with derision, disinterest, or even disdain by prison staff members. The concept of 'spoiled identity' (Goffman 1961) was useful in identifying the ways that older (male) prisoners approached the challenge of identity management. Crawley and Sparks (2006) Crawley (2005) cite the strong stigmas that become attached to sex offenders and older prisoners that overturn a life of respectability. Against this background, individuals possess multiple spoiled identities, older person, a prisoner, and sometimes a sexual offender, which made it difficult for individuals to uphold another identity. The tendency to ignore age-related needs was compounded by the fact that many older prisoners were charged with sexual offences. Although these prisoners often require the same services and provisions as older people outside prison, they constituted an 'undeserving' or 'less eligible' group because of their crimes. What became apparent was the tension between the needs of an older prisoner and his legitimate expectation of moral consideration as an older individual and officers' perception of him as a 'dirty old man'. Similarly, there were times that prison staff attributed wisdom to old age, which might also be regarded as a positive stereotype. However, this did not apply to all prisoners in the sample who, by virtue of their imprisonment in late(r) life, were perceived as unable to live up to this tacit expectation of old age. The research thus confirms the hypothesis that (some of) the needs of this group remained unrecognised or unaddressed as they were considered to 'play make-believe', be 'less deserving', or 'should have known better'.

Prisoners who were regarded as 'old' were not seen as 'proper' prisoners by prison staff due to their compliance, dependency, safer nature, and greater need for care (cf. Crawley, 2005). Much of this is related to long-standing gendered understandings of caring for others that has been primarily linked to women and which was found to pose difficulties in a prison maintained as a 'hypermasculine environment' (Toch, 1998). In Chap. 4, it was argued that caring for older prisoners was not only perceived as contrary to the dignity of men but also confronted some prison professionals with their own finity, making it easier to turn away from, or disparage, them. Staff perceived the frailest individuals as embodiments of pathology rather than ageing people (Baars 2012). Older adults, especially the less able bodied or sound of mind, were likely to be demoted to the status of 'nonperson', often shared as Settersten (2002) suggests, with children to delegitimise their claims to consideration. In addition to this, Chap. 5 illustrates how some older prisoners internalised negative stereotypes associated with old age. Ham's (1976) description of how institutional life for the older population compounds the loss of autonomy with additional pains; it is neatly summed up as one epitomised by:

> the interminable wait for the little and big necessities of survival. This frequent negation of the humanness of the aged is based on the assumption that aged people should be allowed to wait, often unnecessarily, because they have nothing better to do anyway. This waiting and idleness was compounded by the element of captivity. (Ham, 1976: 23)

This book shows the degree to which institutional thoughtlessness is embedded in the prison world (hereinafter 'IT') through spatial and social practices. One of the most significant findings to emerge from this study is that the different dimensions of age strongly affect the validity of the concept of IT. Strongly evident in both the segregated and integrated prison settings, IT was significantly associated with prison staff culture where the dimensions of age played a role. Whilst suitable infrastructure was lacking, many problems could have been overcome with a little more consideration to minor adjustments from prison actors. Thoughtfulness in social practices, however, was mainly demonstrated by older prisoners' prisoner counterparts or by distinct care professionals such as the psychiatric nurse.

The findings also point to some fundamental similarities in the way the social prison environment is shaped across settings. Whilst in the 'integrated' prison setting there was virtually no specific attention for (old) age, the 'segregated' prison setting paid more systemic consideration to a select group of prisoners yet through the bio-medical construct(s) of (old) age. This was epitomised by the figure of the psychiatric nurse who was assigned to the gero-medical unit, the volunteers, and the activities provided by both of these actors. Somewhat similar to care providers who need a 'mandate' of care from their patients, care providers in the second prison setting (though not the first) also needed to be mandated by the institution. With the exception of the psychiatric nurse in the second prison setting, both settings conspicuously failed to address the variability that characterised their older populations. As far as older prisoners' horizontal relations were concerned, it was observed how 'old heads' (Kreager et al. 2017) were able to establish a favourable position within the prison(er) community. Conversely, some older prisoners were found to be subject to victimisation, although this was less than apparent than that by the extant literature: being housed in a specialised gero-medical unit did not indemnify older prisoners against inmate-on-inmate victimisation.

The at times internalised, yet nonetheless dehumanising and objectifying processes, which were extensively documented in Chap. 4, were arguably exacerbated by an ageist prison culture that disparaged care dependency, which means that the human dignity or 'personhood' of some old-age adult prisoners is at risk.

In Chap. 5, we viewed how several prisoners who were already chronologically advanced in age were able to (re)assess their life expectancy or prison time. It was argued that the separation from significant others had a profound adverse effect on older people's lives, whereby 'advanced age' serves as a proxy for 'perceived time left'. Carstensen's (1992) socioemotional selectivity theory (SST) argued that when ageing adults perceive the time that is left as scarce, they will give priority to the fulfilment of present-oriented goals, such as emotional satisfaction and meaning, over typical youth and future-oriented goals, such as social stimulation and gaining information for future use. Although studies investigating the social involvement of captives in later life describe the ageing prison population mainly in terms of limited social interactions and shrinking social networks, the older inmates in this study were found to establish smaller, yet more intimate social networks that increased in emotional value. This resonated with some of the findings of Baumeister and Keller (2011), and Bond et al. (2005), underlining the importance of maintaining close contact with a world that is external to prison. The evidence presented in Chap. 5 demonstrated the importance of such relationships and the struggles that accompanied the maintenance of such relationships for older prisoners. This brings us to a discussion on coping.

## Coping and Meaning in Life

It was argued in Chap. 5 that understanding the prison experience as a stressor demanded taking into account not only the various 'dimensions of age' but also the notion of 'coping'—a process which comprises *changing the meaning* of the stressor and/or transferring attention away from the stressor. It was found that external coping structures, such as family and prison activities, were limited. Even if infrequently accessible within the constraints of the prison environment, the older prisoner is still largely dependent on access to and from the outside world and to tangible resources. For this reason, older prisoners were found to call upon resources within the person in response to their particular stressors. Older prisoners mainly drew upon intrapsychic processes, which enabled them to tap into previous life experience(s) with the benefit of a higher emotional age. In the same vein, it was observed that older prisoners drew primarily upon emotion-focused coping, in contrast to problem-focused strategies. Whilst this might be explained by inherent features of the prison environment—it is often the only resource remaining in prison—some 'old-age prisoners' were found to be disregarded within prison as a result of ageist manifestations.

It was put forward in Chap. 5 that the disruption caused by imprisonment is most profound for those prisoners approaching midlife (Uggen et al. 2005). Despite encountering various stressors of imprisonment, the older prisoners in our study tended to be relatively well-placed to cope with imprisonment. The predominance of 'lower-level strategies' suggested that older adults had the ability to survive, whilst generativity provided scope for thinking about how older prisoners could thrive within a prison context. It is suggested that generativity is one potential path to reveal the nature of how older prisoners can achieve meaning in life. Through various encounters, it is shown how older prisoners attempt to retain at least some degree of individuality via generativity.

These examples have underpinned the assertion that older prisoners still have important developmental tasks and roles to fulfil. The plethora of roles and activities that older prisoners were, and sought to be, involved with showed the desire on their part to contribute to the development of other people and to the life of more than one generation. Based on these findings, it is strongly arguable that imprisonment tends to impede generative behaviour by discarding, challenging, or even removing opportunities for generative manifestations. It is, however, an opportunity missed for prisons to demonstrate *thoughtfulness* towards (older) prisoners and to realise the significant untapped potential for generativity that older prisoners possess. The constraints within the prison system highlighted throughout this research mean that these attributes remain unacknowledged, unrecognised, and unaddressed.

## Characteristics of Ageing, Imprisonment, and Ageing in Prison

So, is it possible to say what is distinctive about (the experiences of) the older prisoner? This is best illustrated by considering the distinctiveness of the experiences of (a) old-age adults, (b) prisoners, and (c) old-age prisoners.

Growing older has been understood throughout as a process of both increasing vulnerability on the one hand and the development of a deepening uniqueness and expanding wealth of experience on the other. The vulnerability associated with growing older was considered in terms of advancing ageing processes such as the increasing susceptibility to disease, physical deterioration, and the increasing likelihood to encounter risks. All these are inextricably bound to the *human condition* as a result of an extended lifespan: as we age, we will experience mental injuries also: disappointment, loss, bereavement, and so on. Drawing on Baars (2012), who viewed growing older in terms of deepening uniqueness and the simultaneous growth of a wealth of differentiated experiences, the findings suggest that this can be understood more in terms of increasing self-referentiality and greater individuality, implicit in a process of increasing differentiation of the life story. With ageing, this complexity increases, which contrasts sharply with generalisations about older adults that tend to pin them down on a distinct characteristic, such as chronological age or disease. Growing older carries with it the potential of achieving a greater sensitivity towards individual differences and the nuanced changes that occur during a long life. Consequently, older persons interact with the vulnerability inherent to human existence and are uniquely placed to pass their individuality and experiences on to others.

The psychological effects of imprisonment on prisoners have been well documented in previous chapters (Sykes 2007; Goffman 1961) and argued to be an experience of mental injury to its captives. That given, it follows that the personhood of *all* captives is denied, including inter alia their personal life stories and variability in ageing, which expands the validity of institutional thoughtlessness to a wider realm of moral considerability.

In both settings, the intersectionality between old age and prisoner status meant that older prisoners are adversely affected by the dual experience of ageing and imprisonment. The vulnerabilities associated with growing older were compounded, whilst the strengths of growing older are simultaneously disregarded or undermined by mechanisms that were argued to be inherent to the prison experience. Nevertheless, much evidence was found to show that older prisoners, through coping strategies and processes of meaning giving, drew upon the advantage of growing older in terms of their deepening wealth of experience and uniqueness. This reflected an interplay between the lives and experiences of prisoners prior to their admission into prison and what occurred during imprisonment, both in terms of ageing and incarceration.

Throughout this research, the concept of 'institutional thoughtlessness' (IT) has been interpreted through the lenses of sameness and otherness towards prisoners growing older. IT was observed when equal treatment was applied to unequal cases (sameness) and when equal cases were treated unequally (otherness)—in both instances, ageism militated against the principle of fair treatment on grounds of age. The findings thus gave rise to question the 'thoughtlessness of institutional thoughtlessness'. It might be argued that the concept of 'institutional thoughtlessness' in the current scientific view is to some extent also 'thoughtless' as it fails to fully recognise ageing as an existential process, the different dimensions of age, and the ambiguity and heterogeneity of experiences in later life. The way in which this life stage should be grasped is both characterised by a growing vulnerability and an increasing wealth of experience. Thus, it is further argued that there is a compelling case for revaluating the process of growing older with regard to group(s) of prisoners that are fundamentally undervalued. Unsurprisingly, in studying the distinctiveness of imprisonment in later life, its non-distinctiveness has also surfaced in a number of findings which show similarities to prisoners in general.

Although the prison setting can hardly be described as an enriched environment, the answer to ageing is neither to place them outside of society nor to put them between brackets. It is important for older prisoners to live in an enriched environment, to be actively involved, to be able to take responsibility, and to have an opportunity to engage in activities that are individually adapted. In addition, the affirmation of past identities needs to be made rewarding, rather than treated as something which ought to be ignored or undermined. In partial accordance with Baars (2012), the findings in both settings extend her idea that a prison culture of dealing with ageing prisoners cannot solely exist of medical interventions but must also take into account that growing older is also an existential process. In short, prisons need to employ non-bio-medical constructs associated with ageing if they are going to be able to facilitate the inclusion of old-age prisoners living with disease and/or disability (Carver and Buchanan 2016).

## Critical Reflections

This work, like any other research project, corresponds to the adage that 'choosing means losing'. Some of the cognate research opportunities, encountered in the literature or the field, that might be considered underexplored but not considered here include the importance of resettlement and release of old-age prisoners in understanding meaning in imprisonment in later life and although female prisoners were part of the sample, a more rigorous interrogation of gender and ageing in prison. The construction of the category of the 'older prisoner' also prompts a more investigation. How this group is constructed might well determine how their experiences are perceived, responses gleaned, and how any suggestions over how the lives of prisoners in later life may be reshaped are formulated. I am thus mindful that the criteria used here also form the subject of scrutiny. In that respect, like any serious scientific enquiry, the results ought to be considered in their complexity and entirety and as a contribution to the continuation of a 'socially constructed' category: the older prisoner.

## References


Appelrouth, S., & Edles, L. D. (2008). *Classical and Contemporary Sociological Theory: Text and Readings*. Pine Forge Press.

Baars, J. (2012). Het nieuwe ouder worden. Paradoxen en perspectieven van leven in de tijd. (3de dr.). SWP Uitgeverij.

Baumeister, B., & Keller, S. (2011). Alt werden im Straf- und Massnahmenvollzug. ZHAW Soziale Arbeit.

Bond, G. D., Thompson, L. A., & Malloy, D. M. (2005). Lifespan Differences in the Social Networks of Prison Inmates. *International Journal of Aging & Human Development, 61*(3), 161–178. https://doi.org/10.2190/7H5P-2AHJ-L34Q-GW9U.

Carstensen, L. L. (1992). Social and Emotional Patterns in Adulthood: Support for Socioemotional Selectivity Theory. *Psychology and Aging, 7*(3), 331–338.

Carver, L. F., & Buchanan, D. (2016). Successful Aging: Considering Non-biomedical Constructs. *Clinical Interventions in Aging, 11*, 1623–1630. https://doi.org/10.2147/CIA.S117202.

Crawley, E. (2005). Institutional Thoughtlessness in Prisons and its Impacts on the Day-to-Day Lives of Elderly Men. *Journal of Contemporary Justice, 21*(4), 350–363.

Crawley, E., & Sparks, R. (2006). Is There Life After Imprisonment? How Elderly Men Talk about Imprisonment and Release. *Criminology and Criminal Justice, 6*(1), 63–82. https://doi.org/10.1177/1748895806060667.

De Motte, C. (2015). *Understanding Older Male Prisoners' Satisfaction with Quality of Life and Wellbeing*. Nottingham Trent University. Retrieved from http://irep.ntu.ac.uk/id/eprint/31213/1/Claire%20de%20Motte%202015%20excl3rdpartycopright.pdf.

Derkx, P. H. J. M. (2011). *Humanisme, zinvol leven en nooit meer 'ouder worden': Een levensbeschouwelijke visie op ingrijpende biomedisch-technologische levensverlenging*. VUBPress.

Fenton, S.-J., & Draper, H. (2014). *The Experience of Ageing Is a Complex One That Is Only Loosely Associated with Number of Years Lived*. Birmingham Policy Commission. Retrieved from https://www.birmingham.ac.uk/Documents/research/policycommission/healthy-ageing/1-The-experience-of-ageing-is-complex-updated.pdf.

Ferraro, K. F. (2013). The Gerontological Imagination. In J. M. Wilmoth & K. F. Ferraro (Eds.), *Gerontology: Perspectives and Issues* (4th ed., pp. 1–18). Springer Publishing Company.

Foucault, M. (1984). Des Espace Autres. Heterotopies. (J. Miskowiec, Transl.). *Architecture, Mouvement, Continuité, 5*, 46–49.

Giddens, A. (1984). *The Constitution of Society*. Cambridge: Polity Press.

Godfrey, M., Townsend, J., & Denby, T. (2004). *Building a Good Life for Older People in Communities: The Experience of Ageing in Time and Place*. Joseph Rowntree Foundation.

Goetting, A. (1985). Racism, Sexism, and Ageism in the Prison Community. *Federal Probation, 44*(3), 10–22.

Goffman, E. (1961). *Asylums: Essays on the Social Situation of Mental Patients and Other Inmates*. Anchor Books.

Goulding, P. (2013). Silver Bullet or Confused Greying Fox? Best Practice Support Model for Older Prisoners. Wintringham Specialist Aged Care. Retrieved from http://www.wintringham.org.au/file/434/I/Best_Practice_Support_Model_for_Older_Prisoners.pdf.

Ham, J. N. (1976). *The Forgotten Minority: An Exploration of Long-term Institutionalized Aged and Aging Male Prison Inmates*. University of Michigan.

Kreager, D. A., Young, T. N. J., Haynie, D., Bouchard, M., Schaefer, D., & Zajac, G. (2017). Where "Old Heads" Prevail: Inmate Hierarchy in a Men's Prison Unit. *American Sociological Review, 82*, 1–34. https://doi.org/10.1177/0003122417710462.

Mann, N. (2008). *Doing Harder Time? The Experience of An Ageing Male Prison Population in England and Wales*. University of Essex.

Mann, N. (2012). *Doing Harder Time? The Experiences of an Ageing Male Prison Population in England and Wales*. Ashgate.

Mann, N. (2016). Older Age, Harder Time: Ageing and Imprisonment. In Y. Jewkes, B. Crewe, & J. Bennett (Eds.), *Handbook on Prisons* (2nd ed., pp. 514–528). Routledge.

Pinquart, M. (2002). Creating and Maintaining Purpose in Life in Old Age: A Meta-Analysis. *Ageing International, 27*(2), 90–114. https://doi.org/10.1007/s12126-002-1004-2.

Rowe, J. W., & Kahn, R. L. (1987). Human Aging: Usual and Successful. *Science, 237*(4811), 143–149.

Settersten, R. A. (2002). Social Sources of Meaning in Later Life. In R. S. Weiss & S. A. Bass (Eds.), *Challenges of the Third Age: Meaning and Purpose in Later Life*. Oxford University Press.

Sykes, G. M. (1958). *The Society of Captives: A Study of a Maximum Security Prison*. Princeton University Press.

Sykes, G. M. (2007). *The Society of Captives: A Study of a Maximum Security Prison*. Princeton University Press.

Toch, H. (1998). Hypermasculinity and Prison Violence. In L. H. Bowker (Ed.), *Masculinities and Violence*. SAGE Publications.

Uggen, C., Wakefield, S., & Western, B. (2005). Work and Family Perspectives on Reentry. In J. Travis & C. Visher (Eds.), *Prisoner Reentry and Crime in America* (pp. 209–243). Cambridge University Press.

Wahidin, A. (2004). *Older Women in the Criminal Justice System: Running Out of Time*. Jessica Kingsley Publishers.

# Index[1]



[1] Note: Page numbers followed by 'n' refer to notes.



D. Humblet, *The Older Prisoner*, Palgrave Studies in Prisons and Penology,
https://doi.org/10.1007/978-3-030-60120-1

The manufacturer's authorised representative in the EU is Springer
Nature Customer Service Centre GmbH, Europaplatz 3, 69115 Heidelberg,
Germany. If you have any concerns regarding our products, please
contact ProductSafety@springernature.com

Printed and bound by CPI Group (UK) Ltd, Croydon, CR0 4YY
30/09/2026
02225696-0002